Spherical Harmonics
in p Dimensions

Spherical Harmonics in p Dimensions

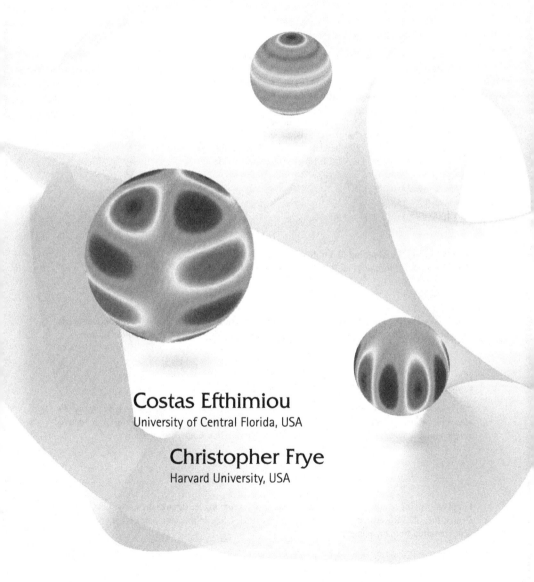

Costas Efthimiou
University of Central Florida, USA

Christopher Frye
Harvard University, USA

World Scientific

NEW JERSEY · LONDON · SINGAPORE · BEIJING · SHANGHAI · HONG KONG · TAIPEI · CHENNAI

Published by

World Scientific Publishing Co. Pte. Ltd.

5 Toh Tuck Link, Singapore 596224

USA office: 27 Warren Street, Suite 401-402, Hackensack, NJ 07601

UK office: 57 Shelton Street, Covent Garden, London WC2H 9HE

Library of Congress Cataloging-in-Publication Data
Efthimiou, Costas, author.
    Spherical harmonics in p dimensions / by Costas Efthimiou (University of Central Florida,
USA) & Christopher Frye (Harvard University, USA).
        pages cm
    Includes bibliographical references and index.
    ISBN 978-9814596695 (hardcover : alk. paper)
    1. Spherical harmonics. 2. Spherical functions. 3. Legendre's polynomials.
4. Mathematical physics. I. Frye, Christopher, author. II. Title.
    QC20.7.S645E38 2014
    515'.785--dc23

    2014000203

British Library Cataloguing-in-Publication Data
A catalogue record for this book is available from the British Library.

Cover image: 3 spherical balls on cover: Adapted from 'Spherical harmonics in 3D', Wikipedia Commons. This file is licensed under the Creative Commons Attribution-Share Alike 3.0 Unported license.

Printed in Singapore

Contents

Preface   vii

Acknowledgements   ix

List of Symbols   xi

1 Introduction and Motivation   1
   1.1 Separation of Variables   2
   1.2 Quantum Mechanical Angular Momentum   9

2 Working in p Dimensions   13
   2.1 Rotations in $\mathbb{E}^p$ 14
   2.2 Spherical Coordinates in p Dimensions   15
   2.3 The Sphere in Higher Dimensions   20
   2.4 Arc Length in Spherical Coordinates   24
   2.5 The Divergence Theorem in $\mathbb{E}^p$ 26
   2.6 Δ_p in Spherical Coordinates   30
   2.7 Problems   34

3 Orthogonal Polynomials   39
   3.1 Orthogonality and Expansions   39
   3.2 The Recurrence Formula   42
   3.3 The Rodrigues Formula   45
   3.4 Approximations by Polynomials   48
   3.5 Hilbert Space and Completeness   57
   3.6 Problems   62

4 Spherical Harmonics in p Dimensions   **63**
   4.1   Harmonic Homogeneous Polynomials 63
   4.2   Spherical Harmonics and Orthogonality 71
   4.3   Legendre Polynomials 78
   4.4   Boundary Value Problems 105
   4.5   Problems 116

5 Solutions to Problems   **119**
   5.1   Solutions to Problems of Chapter 2 119
   5.2   Solutions to Problems of Chapter 3 127
   5.3   Solutions to Problems of Chapter 4 131

Bibliography   **139**

Index   **141**

Preface

We prepared the following book in order to make several useful topics from the theory of special functions, in particular the spherical harmonics and Legendre polynomials of $\mathbb{R}^p$, available to undergraduates studying physics or mathematics. With this audience in mind, nearly all details of the calculations and proofs are written out, and extensive background material is covered before beginning the main subject matter. The reader is assumed to have knowledge of multivariable calculus and linear algebra (especially inner product spaces) as well as some level of comfort with reading proofs.

Literature in this area is scant, and for the undergraduate it is virtually nonexistent. To find the development of the spherical harmonics that arise in $\mathbb{R}^3$, physics students can look in almost any text on mathematical methods, electrodynamics, or quantum mechanics (see [3], [5], [10], [12], [17], for example), and math students can search any book on boundary value problems, PDEs, or special functions (see [7], [20], for example). However, the undergraduate will have a very difficult time finding *accessible* material on the corresponding topics in arbitrary $\mathbb{R}^p$.

We have been greatly influenced by Hochstadt's *The Functions of Mathematical Physics* [9]. When this book was prepared and released as notes, Hochstadt's book was out of print. Fortunately, Dover reprinted it in the Spring of 2012. However, this book contains a more expanded and detailed point of view than in Hochstadt's book, as well additional information and a chapter with the solutions to all problems. There are several additional references (see [2], [6], [13], [21], [22], for example) where the reader can search for information on the topic of this book, but either the coverage is brief or the level of difficulty is considerably higher. In addition, the point of view is very different from the one adopted in this work. Therefore, we hope that the current book will become a useful supplement for any reader interested in special functions and mathematical physics, especially students who learn the topic.

Acknowledgements

Christopher Frye worked on this book as an undergraduate and is very grateful to Costas Efthimiou for his guidance and assistance during this project. Chris would also like to thank Professors Maxim Zinchenko and Alexander Katsevich for their valuable comments and suggestions on the presentation of the ideas. In addition, he thanks his friends Jie Liang, Byron Conley, and Brent Perreault for their feedback after proofreading portions of this manuscript. He is also grateful to Kyle Anderson for assistance with one of the figures.

This work has been supported in part by a National Science Foundation Award DUE 0963146 as well as UCF SMART and RAMP Awards. Chris Frye is grateful to the Burnett Honors College (BHC), the Office of Undergraduate Research (OUR), and the Research and Mentoring Program (RAMP) at UCF for their generous support during his studies. He would like to thank especially Alvin Wang, Dean of BHC; Martin Dupuis, Assistant Dean of BHC; Kim Schneider, Director of OUR; and Michael Aldarondo-Jeffries, Director of RAMP. Last but not least, he thanks Paul Steidle for providing him a quiet office to work in while writing.

List of Symbols

$f : X \to Y$	a function f mapping set X into set Y
$\mathbb{R}^p$	the p-dimensional vector space of real p-tubles
$\mathbb{E}^p$	the p-dimensional Euclidean space
$x \in \mathbb{R}^p$	point x in $\mathbb{R}^p$, written without vector arrow
$|x|$	distance from point x to the origin of $\mathbb{E}^p$
$x_1, \ldots, x_p$	Cartesian coordinates in $\mathbb{R}^p$
$r, \phi, \theta_1, \ldots, \theta_{p-2}$	Spherical coordinates in $\mathbb{R}^p$
∇_p	del operator in $\mathbb{E}^p$
Δ_p	Laplace operator in $\mathbb{E}^p$
$\Delta_{S^{p-1}}$	Laplace-Beltrami operator in S^{p-1}
M^t	transpose of matrix M
ξ, η	unit vectors in $\mathbb{E}^p$
$S_\delta^{p-1}(x_0)$	sphere of radius δ and center x_0 in $\mathbb{E}^p$
S^{p-1}	unit sphere in $\mathbb{E}^p$
$A_{p-1}(\delta)$	surface area of S_δ^{p-1}
Ω_{p-1}	surface area of S^{p-1}
$B_\delta^{p-1}(x_0)$	open ball of radius δ and center x_0 in $\mathbb{E}^p$
B^p	unit ball in $\mathbb{E}^p$
$V_p(\delta)$	volume of B_δ^p

$\Gamma(t)$	gamma function
$\vec{F}$	fixed vector or vector field in $\mathbb{R}^p$ (only occasionally used)
$\hat{n}$	unit normal vector (only occasionally used)
$d^p x$	differential volume element in $\mathbb{E}^p$
$d\sigma$	differential surface element
$\mathcal{P}$	set of real polynomials in one variable
$\langle \cdot, \cdot \rangle$	inner product
w	weight function in inner product
$\|x\|$	norm of vector x
δ_{ij}	Kronecker delta
$(k)_\ell$	falling factorial
$\binom{n}{k}$	n choose k
$B_n(x; f)$	n-th Bernstein polynomial of f
$\text{Prob}(A)$	probability for event A
L.O.T.	lower order term
$K(p, n)$	number of n-th degree linearly independent homogeneous polynomials in $\mathbb{R}^p$
$N(p, n)$	number of n-th degree linearly independent homogeneous harmonic polynomials in $\mathbb{R}^p$
$H_n(x)$	homogeneous polynomial of degree n
$Y_{n,j}(\xi)$	j-th spherical harmonic of order n
$L_n(\xi)$	n-th degree Legendre polynomial
$P_{n,p}(t)$	n-th degree Legendre polynomial in $\mathbb{E}^p$ (p usually omitted)
$P_n^j(t)$	associated Legendre polynomial
$G(x)$	Green's function
$\delta(x - x_0)$	Dirac delta function

Chapter 1

Introduction and Motivation

Many important equations in physics involve the *Laplace operator*, which is given by

$$\Delta_2 = \frac{\partial^2}{\partial x^2} + \frac{\partial^2}{\partial y^2} \,, \tag{1.1}$$

$$\Delta_3 = \frac{\partial^2}{\partial x^2} + \frac{\partial^2}{\partial y^2} + \frac{\partial^2}{\partial z^2} \,, \tag{1.2}$$

in two and three dimensions[1], respectively. We will see later (Proposition 2.1) that the Laplace operator is invariant under a rotation of the coordinate system. Thus, it arises in many physical situations in which there exists spherical symmetry, i.e., where physical quantities depend only on the radial distance r from some center of symmetry $\mathcal{O}$. For example, the electric potential V in free space is found by solving the *Laplace equation*,

$$\Delta\Phi = 0 \,, \tag{1.3}$$

which is rotationally invariant. Also, in quantum mechanics, the wave function ψ of a particle in a central field can be found by solving

---

[1]We may drop the subscript if we want to keep the number of dimensions arbitrary.

1

the *time-independent Schrödinger equation,*

$$\left[-\frac{\hbar^2}{2m}\Delta + V(r) \right] \psi \;=\; E\psi \tag{1.4}$$

where $\hbar$ is Planck's constant, m is the mass of the particle, $V(r)$ is its potential energy, and E is its total energy.

We will give a brief introduction to these problems in two and three dimensions to motivate the main subject of this discussion. In doing this, we will get a preview of some of the properties of *spherical harmonics* — which, for now, we can just think of as some special set of functions — that we will develop later in the general setting of $\mathbb{R}^p$.

1.1   Separation of Variables

Two-Dimensional Case

Since we are interested in problems with spherical symmetry, let us rewrite the Laplace operator in spherical coordinates, which in $\mathbb{R}^2$ are just the ordinary polar coordinates[2],

$$r = \sqrt{x^2 + y^2}, \quad \phi = \tan^{-1}\left(\frac{y}{x}\right). \tag{1.5}$$

Alternatively,

$$x = r\cos\phi, \quad y = r\sin\phi.$$

Using the chain rule, we can rewrite the Laplace operator as

$$\Delta_2 = \frac{\partial^2}{\partial r^2} + \frac{1}{r}\frac{\partial}{\partial r} + \frac{1}{r^2}\frac{\partial^2}{\partial \phi^2}. \tag{1.6}$$

In checking this result, perhaps it is easiest to begin with (1.6) and recover (1.1). First we compute

$$\frac{\partial}{\partial r} = \frac{\partial x}{\partial r}\frac{\partial}{\partial x} + \frac{\partial y}{\partial r}\frac{\partial}{\partial y} = \cos\phi\frac{\partial}{\partial x} + \sin\phi\frac{\partial}{\partial y}, \tag{1.7}$$

---

[2]The polar angle ϕ actually requires a more elaborate definition, since $\tan^{-1}$ only produces angles in the first and fourth quadrants. However, this detail will not concern us here.

which implies that

$$\frac{\partial^2}{\partial r^2} = \cos^2\phi\frac{\partial^2}{\partial x^2} + 2\sin\phi\cos\phi\frac{\partial^2}{\partial x\partial y} + \sin^2\phi\frac{\partial^2}{\partial y^2}\,,$$

and

$$\frac{\partial}{\partial\phi} = \frac{\partial x}{\partial\phi}\frac{\partial}{\partial x} + \frac{\partial y}{\partial\phi}\frac{\partial}{\partial y} = -r\sin\phi\frac{\partial}{\partial x} + r\cos\phi\frac{\partial}{\partial y}\,,$$

which gives

$$\frac{\partial^2}{\partial\phi^2} = r^2\sin^2\phi\frac{\partial^2}{\partial x^2} - 2r^2\sin\phi\cos\phi\frac{\partial^2}{\partial x\partial y} + r^2\cos^2\phi\frac{\partial^2}{\partial y^2}\,.$$

Inserting these into (1.6) gives us back (1.1).

Thus, (1.3) becomes

$$\frac{\partial^2\Phi}{\partial r^2} + \frac{1}{r}\frac{\partial\Phi}{\partial r} + \frac{1}{r^2}\frac{\partial^2\Phi}{\partial\phi^2} = 0\,.$$

To solve this equation it is standard to assume that

$$\Phi(r,\phi) = \chi(r)\,Y(\phi)\,,$$

where $\chi(r)$ is a function of r alone and $Y(\phi)$ is a function of ϕ alone. Then

$$Y\frac{d^2\chi}{dr^2} + \frac{Y}{r}\frac{d\chi}{dr} + \frac{\chi}{r^2}\frac{d^2Y}{d\phi^2} = 0\,.$$

Multiplying by $r^2/\chi Y$ and rearranging,

$$\frac{r^2}{\chi}\frac{d^2\chi}{dr^2} + \frac{r}{\chi}\frac{d\chi}{dr} = \frac{-1}{Y}\frac{d^2Y}{d\phi^2}\,.$$

We see that a function of r alone (the left side) is equal to a function of ϕ alone (the right side). Since we can vary r without changing ϕ, i.e., without changing the right side of the above equation, it must be that the left side of the above equation does not vary with r either. This means the left side of the above equation is not really a function of r but a constant. As a consequence, the right side is the same constant. Thus, for some $-\lambda$ we can write

$$\frac{-1}{Y}\frac{d^2Y}{d\phi^2} = -\lambda = \frac{r^2}{\chi}\frac{d^2\chi}{dr^2} + \frac{r}{\chi}\frac{d\chi}{dr}. \tag{1.8}$$

We solve $Y'' = \lambda Y$ to get the linearly independent solutions

$$Y(\phi) = \begin{cases} e^{\sqrt{\lambda}\phi},\ e^{-\sqrt{\lambda}\phi} & \text{if } \lambda > 0, \\ 1,\ \phi & \text{if } \lambda = 0, \\ \sin\left(\sqrt{|\lambda|}\phi\right),\ \cos\left(\sqrt{|\lambda|}\phi\right) & \text{if } \lambda < 0, \end{cases}$$

but we must reject some of these solutions. Since (r_0, ϕ_0) represents the same point as $(r_0, \phi_0 + 2\pi k)$ for any $k \in \mathbb{Z}$, we require $Y(\phi)$ to have period 2π. Thus, we can only accept the linearly independent periodic solutions 1, $\sin\left(\sqrt{|\lambda|}\phi\right)$, and $\cos\left(\sqrt{|\lambda|}\phi\right)$, where $\sqrt{|\lambda|}$ must be an integer. Then, let us replace λ with $-m^2$ and write our linearly independent solutions to $Y'' = -m^2 Y$ as

$$Y_{1,n} = \cos(n\phi), \qquad Y_{2,m} = \sin(m\phi), \tag{1.9}$$

where[3] $n \in \mathbb{N}_0$ and $m \in \mathbb{N}$. Notice from (1.6) that $\partial^2/\partial\phi^2$ is the angular part of the Laplace operator in two dimensions and that the solutions given in (1.9) are eigenfunctions of the $\partial^2/\partial\phi^2$ operator. We will see in Chapter 4 that the functions in (1.9) are actually spherical harmonics; however, since we have not yet given a definition of a spherical harmonic, for now we will just refer to these as functions Y. The reader should keep in mind that characteristics of the Y's we comment on here will generalize when we move to $\mathbb{R}^p$.

We can also solve easily for the functions $\chi(r)$ that satisfy (1.8), but this does not concern us here. Let us instead notice a few properties of the functions Y. Let us consider these to be functions $r^m \sin(m\phi)$, $r^n \cos(n\phi)$ on $\mathbb{R}^2$ that have been restricted to the unit circle, where $r = 1$, and let us analyze the extended functions on $\mathbb{R}^2$.

---

[3] We use the notation $\mathbb{N} = \{1, 2, \dots\}$ and $\mathbb{N}_0 = \{0, 1, 2, \dots\}$.

We will first rewrite them using *Euler's formula*, $e^{i\phi} = \cos\phi + i\sin\phi$, which implies

$$(x + iy)^n = \left(re^{i\phi}\right)^n = r^n e^{in\phi} = r^n\left[\cos\left(n\phi\right) + i\sin\left(n\phi\right)\right],$$

$$(x - iy)^n = \left(re^{-i\phi}\right)^n = r^n e^{-in\phi} = r^n\left[\cos\left(n\phi\right) - i\sin\left(n\phi\right)\right],$$

so that

$$r^n \cos\left(n\phi\right) = \frac{1}{2}\left[(x + iy)^n + (x - iy)^n\right] \stackrel{\text{def}}{=} H_{1,n}(x, y),$$

$$r^n \sin\left(n\phi\right) = \frac{1}{2i}\left[(x + iy)^n - (x - iy)^n\right] \stackrel{\text{def}}{=} H_{2,n}(x, y).$$

We notice that the Y's can be written as polynomials restricted to the unit circle, where $r = 1$. Furthermore, observe that

$$H_{1,n}(tx, ty) = t^n H_{1,n}(x, y), \quad \text{and} \quad H_{2,n}(tx, ty) = t^n H_{2,n}(x, y);$$

we call polynomials with this property *homogeneous of degree n*. Moreover, the reader can also check that these polynomials satisfy the Laplace equation (1.3), by either using (1.1) or the Laplace operator in polar coordinates (1.6) for the computation, i.e.,

$$\Delta_2 H_{1,n} = 0, \quad \text{and} \quad \Delta_2 H_{2,n} = 0.$$

Let us also notice that the Y's of different degree are *orthogonal over the unit circle*, which means

$$\int_0^{2\pi} \sin\left(n\phi\right)\sin\left(m\phi\right) d\phi = 0, \quad \text{if } n \neq m,$$

$$\int_0^{2\pi} \cos\left(n\phi\right)\cos\left(m\phi\right) d\phi = 0, \quad \text{if } n \neq m,$$

$$\int_0^{2\pi} \sin\left(n\phi\right)\cos\left(m\phi\right) d\phi = 0, \quad \text{if } n \neq m,$$

as we can easily compute by taking advantage of Euler's formula. For instance, we can calculate

$$\int_0^{2\pi} \sin(n\phi)\sin(m\phi)\, d\phi = \int_0^{2\pi} \frac{e^{in\phi} - e^{-in\phi}}{2} \cdot \frac{e^{im\phi} - e^{-im\phi}}{2}\, d\phi,$$

which becomes

$$\int\limits_0^{2\pi} \left(\frac{e^{i(m+n)\phi} - e^{-i(m+n)\phi}}{2} - \frac{e^{i(m-n)\phi} + e^{-i(m-n)\phi}}{2} \right) d\phi \,,$$

or

$$\int\limits_0^{2\pi} \Big(\sin[(m+n)\phi] - \cos[(m-n)\phi] \Big)\, d\phi = 0\,,$$

for $n \neq m$. The reader can check the rest in a similar fashion.

Finally, by recalling the theorems of Fourier analysis[4], we know that any "reasonable" function defined on the unit circle can be expanded in a *Fourier series*. That is, given a function $f : [0, 2\pi) \to \mathbb{R}$ satisfying certain conditions (that do not concern us in this introduction), we can write

$$f(\phi) = \sum_{m=1}^{\infty} a_m \sin(m\phi) + \sum_{n=0}^{\infty} b_n \cos(n\phi)\,,$$

for some constants a_m, b_n. We say that the Y's make up a *complete set* of functions over the unit circle, since we can expand any nice function $f(\phi)$ defined on $[0, 2\pi)$ in terms of them.

Before we move on, we will show that (1.4) can be approached using a method almost identical to the method we used above. Let us assume that the solution ψ to (1.4) can be written as $\psi(r, \phi) = \chi(r)Y(\phi)$. Then, using polar coordinates, the equation becomes

$$Y \left(-\frac{\hbar^2}{2m} \frac{d^2}{dr^2} - \frac{\hbar^2}{2mr} \frac{d}{dr} + V(r) \right) \chi + \frac{\chi}{r^2} \left(-\frac{\hbar^2}{2m} \frac{d^2}{d\phi^2} \right) Y = E\chi Y\,.$$

Multiplying by $r^2/\chi Y$ and rearranging,

$$\frac{1}{\chi} \left(-\frac{r^2 \hbar^2}{2m} \frac{d^2}{dr^2} - \frac{r \hbar^2}{2m} \frac{d}{dr} + r^2 V(r) \right) \chi - Er^2 = \frac{1}{Y} \left(\frac{\hbar^2}{2m} \frac{d^2}{d\phi^2} \right) Y\,.$$

---

[4]Doing so will not be necessary to understand the material we present here, but the reader unfamiliar with Fourier analysis may choose to consult [7].

Once again, we see that a function of r alone is equal to a function of ϕ alone, and we conclude that both sides of the above equation must be equal to the same constant. Using the same reasoning as before, we write this constant as $-\ell^2 \hbar^2/2m$, where ℓ is an integer. If we carry out the calculation, we will see that the functions $Y(\phi)$ are the same ones we found previously in this subsection. We will also find the radial equation

$$\frac{1}{\chi}\left(-\frac{r^2\hbar^2}{2m}\frac{d^2}{dr^2} - \frac{r\hbar^2}{2m}\frac{d}{dr} + r^2 V(r)\right)\chi - Er^2 = -\frac{\ell^2\hbar^2}{2m},$$

which we can rewrite as

$$\left\{-\frac{\hbar^2}{2m}\left(\frac{d^2}{dr^2} - \frac{1}{r}\frac{d}{dr}\right) + \left[V(r) + \frac{\ell^2\hbar^2}{2mr^2}\right]\right\}\chi = E\chi.$$

It is interesting to note that this equation resembles (1.4). If we think of r as our single independent variable and χ as our wave function, we have an effective potential energy

$$V_{\text{eff}}(r) = V(r) + \frac{\ell^2\hbar^2}{2mr^2},$$

where we call the second term the *centrifugal* term. Thus χ represents a fictitious particle that feels an effective force

$$\vec{F}_{\text{eff}} = -\nabla V_{\text{eff}} = -\frac{dV_{\text{eff}}}{dr}\hat{r}.$$

We see that the centrifugal term contributes a force

$$\vec{F}_{\text{centrifugal}} = \frac{\ell^2\hbar^2}{mr^3}\hat{r},$$

pushing the particle away from the center of symmetry.

Three-Dimensional Case

Here, we will follow a procedure almost identical to that of the last subsection, but we will not take the discussion as far in three dimensions. The Y's we will find in three dimensions are more widely used than those in any other number of dimensions; however, a thorough development of the functions in $\mathbb{R}^3$ could take many pages, and this would distract us from our goal of moving to p dimensions. Furthermore, the results that we would find in three dimensions are only special cases of more general theorems we will develop later in the discussion. If the reader is interested in studying the usual spherical harmonics of $\mathbb{R}^3$ in depth, there are a multitude of sources we can recommend, including [3], [5], [7] and [20].

Let us rewrite the Laplace operator in spherical coordinates, which are given by

$$r = \sqrt{x^2 + y^2 + z^2}, \quad \theta = \tan^{-1}\left(\frac{\sqrt{x^2 + y^2}}{z}\right), \quad \phi = \tan^{-1}\left(\frac{y}{x}\right).$$
(1.10)

Alternatively,

$$x = r\sin\theta\cos\phi, \quad y = r\sin\theta\sin\phi, \quad z = r\cos\theta.$$

Using the chain rule, we find

$$\Delta_3 = \frac{1}{r^2}\frac{\partial}{\partial r}\left(r^2\frac{\partial}{\partial r}\right) + \frac{1}{r^2}\left[\frac{1}{\sin\theta}\frac{\partial}{\partial\theta}\left(\sin\theta\frac{\partial}{\partial\theta}\right) + \frac{1}{\sin^2\theta}\frac{\partial^2}{\partial\phi^2}\right].$$
(1.11)

As in the two-dimensional case, it is probably easiest to verify this formula by starting with (1.11) and producing (1.2). The reader should check this result for practice, proceeding exactly as we did beginning in (1.7).

Inserting (1.11) into (1.3) gives

$$\frac{1}{r^2}\frac{\partial}{\partial r}\left(r^2\frac{\partial\Phi}{\partial r}\right) + \frac{1}{r^2}\left[\frac{1}{\sin\theta}\frac{\partial}{\partial\theta}\left(\sin\theta\frac{\partial\Phi}{\partial\theta}\right) + \frac{1}{\sin^2\theta}\frac{\partial^2\Phi}{\partial\phi^2}\right] = 0.$$

Searching for solutions Φ in the form $\Phi(r,\theta,\phi) = \chi(r)Y(\theta,\phi)$, where $\chi(r)$ is a function of r alone and $Y(\theta,\phi)$ is a function of only θ and ϕ, this becomes

$$Y\frac{1}{r^2}\frac{d}{dr}\left(r^2\frac{d}{dr}\right)\chi + \frac{\chi}{r^2}\left[\frac{1}{\sin\theta}\frac{\partial}{\partial\theta}\left(\sin\theta\frac{\partial}{\partial\theta}\right) + \frac{1}{\sin^2\theta}\frac{\partial^2}{\partial\phi^2}\right]Y = 0.$$

Multiplying by $r^2/\chi Y$ and rearranging,

$$\frac{1}{\chi}\frac{d}{dr}\left(r^2\frac{d}{dr}\right)\chi = \frac{-1}{Y}\left[\frac{1}{\sin\theta}\frac{\partial}{\partial\theta}\left(\sin\theta\frac{\partial}{\partial\theta}\right) + \frac{1}{\sin^2\theta}\frac{\partial^2}{\partial\phi^2}\right]Y.$$

Since we have found that a function of r alone is equal to a function of only θ and ϕ, we use the same reasoning as in the previous subsection to conclude that both sides of the above equation must be equal to the same constant, call it $-\lambda$. This implies that

$$\left[\frac{1}{\sin\theta}\frac{\partial}{\partial\theta}\left(\sin\theta\frac{\partial}{\partial\theta}\right) + \frac{1}{\sin^2\theta}\frac{\partial^2}{\partial\phi^2}\right]Y = \lambda Y. \qquad (1.12)$$

At this point, we will stop. It turns out that the Y's which satisfy this equation are actually spherical harmonics. Comparing (1.11) and (1.12), we see that the Y's are eigenfunctions of the angular part of the Laplace operator, just as in two dimensions. If we studied the functions Y further we would also find that, analogously to what we noticed in $\mathbb{R}^2$, the Y's form a complete set over the unit sphere, each Y is a homogeneous polynomial restricted to the unit sphere, and these polynomials satisfy the Laplace equation. However, to develop these results and the many others that exist would require us to study Legendre's equation, Legendre polynomials, and associated Legendre functions, and we will choose to leave such an in-depth analysis to the general case of p dimensions.

We will now move on to see how these functions Y are related to angular momentum in quantum mechanics.

1.2 Quantum Mechanical Angular Momentum

We have seen that spherical harmonics in two and three dimensions relate to the one-dimensional sphere (circle) and the two-dimensional sphere (surface of a regular ball) respectively. In quantum mechanics, rotations of a system are generated by the angular momentum operator. Spherical symmetry means invariance under all such rotations.

Therefore, a relation between the theory of spherical harmonics and the theory of angular momentum is not only expected but is a natural and fundamental result.

Recall that in classical mechanics, the angular momentum of a particle is defined by the cross product

$$\vec{L} = \vec{r} \times \vec{p},$$

where $\vec{p}$ is its linear momentum. To find the quantum mechanical angular momentum operator, we make the substitution $p_i \mapsto -i\hbar\, \partial/\partial x_i$ where $x_1 = x$, $x_2 = y$, and $x_3 = z$. Thus, we see that[5]

$$\hat{\vec{L}} = -i\hbar\, \vec{r} \times \nabla,$$

where

$$\nabla = \left(\frac{\partial}{\partial x}, \frac{\partial}{\partial y}, \frac{\partial}{\partial z} \right).$$

Two-Dimensional Case

In the plane, the angular momentum operator has only one component, given by

$$\hat{L} = -i\hbar \left(x\frac{\partial}{\partial y} - y\frac{\partial}{\partial x} \right).$$

Using the polar coordinates defined in (1.5) and the chain rule, we can rewrite this as

$$\hat{L} = -i\hbar\frac{\partial}{\partial \phi},$$

as the reader should have no problem checking using the same strategy s/he used to verify (1.6) and (1.11). Then

$$\hat{L}^2 = -\hbar^2 \frac{\partial^2}{\partial \phi^2},$$

---

[5]The hat above the angular momentum indicates that it is an operator (not a unit vector) in this section.

and we check using (1.9) that the functions $Y(\phi)$ are eigenfunctions of the $\hat{L}^2$ operator. In particular,

$$\hat{L}^2 Y_{m,j}(\phi) = -\hbar^2 \frac{\partial^2}{\partial \phi^2} Y_{m,j} = \hbar^2 m^2 Y_{m,j},$$

and we see that the function $Y_{m,j}$ is associated with the eigenvalue $\hbar^2 m^2$.

In quantum mechanics, operators such as $\hat{L}$ represent dynamical variables. If an operator $\hat{O}$ has eigenfunctions ψ_k with corresponding eigenvalues λ_k, then a particle in state ψ_k will be observed to have a value of λ_k for the dynamical variable $\hat{O}$. Therefore, we see that a particle in the state $Y_{m,j}$ will be observed to have a value of $\hbar^2 m^2$ for its angular momentum squared. We say the function $Y_{m,j}$ carries angular momentum $\hbar m$.

Three-Dimensional Case

Things work similarly in three dimensions, where the angular momentum operator has the components

$$\hat{L}_x = -i\hbar \left(y \frac{\partial}{\partial z} - z \frac{\partial}{\partial y} \right),$$

$$\hat{L}_y = -i\hbar \left(z \frac{\partial}{\partial x} - x \frac{\partial}{\partial z} \right),$$

$$\hat{L}_z = -i\hbar \left(x \frac{\partial}{\partial y} - y \frac{\partial}{\partial x} \right).$$

Using the spherical coordinates defined in (1.10) and the chain rule, we can rewrite these as

$$\hat{L}_x = i\hbar \left(\sin \theta \frac{\partial}{\partial \theta} + \cos \phi \cot \theta \frac{\partial}{\partial \phi} \right),$$

$$\hat{L}_y = -i\hbar \left(\cos \theta \frac{\partial}{\partial \theta} + \sin \phi \cot \theta \frac{\partial}{\partial \phi} \right),$$

$$\hat{L}_z = -i\hbar \frac{\partial}{\partial \phi};$$

the reader should check these formulas. These equations allow us to compute $\hat{\vec{L}}^2 = \hat{\vec{L}} \cdot \hat{\vec{L}} = \hat{L}_x^2 + \hat{L}_y^2 + \hat{L}_z^2$. Carrying out the multiplication,

$$\hat{\vec{L}}^2 = -\hbar^2 \left[\frac{1}{\sin\theta} \frac{\partial}{\partial\theta} \left(\sin\theta \frac{\partial}{\partial\theta} \right) + \frac{1}{\sin^2\theta} \frac{\partial^2}{\partial\phi^2} \right], \tag{1.13}$$

and we see that by (1.12), the functions Y are eigenfunctions of the $\hat{\vec{L}}^2$ operator.

We claimed in the last section that the functions $Y(\theta,\phi)$ were homogeneous polynomials with restricted domain, so let us write $Y_\ell(\theta,\phi)$ where ℓ denotes the degree of homogeneity. In Section 4.2, we will see an easy way to compute the eigenvalue of $\hat{\vec{L}}^2$ associated with Y_ℓ, and it will turn out to be $\hbar^2\ell(\ell+1)$. So we claim that in three dimensions, the function $Y_\ell(\theta,\phi)$ carries an angular momentum of $\hbar\sqrt{\ell(\ell+1)}$.

In Chapter 4, we will give rigorous foundations to the seemingly coincidental facts we have discovered in this chapter about the functions Y that arose as solutions to certain differential equations. But first, we will devote a chapter to gaining some practice and intuition working in $\mathbb{R}^p$.

Chapter 2

Working in p Dimensions

In this chapter, we spend some time developing our skills in performing calculations in the p-dimensional *Euclidean space*[1] $\mathbb{E}^p$, i.e. the vector space of p-tuples of real numbers endowed with the *Euclidean metric*, and exercising our abilities in visualizing a p-dimensional space for arbitrary natural number p. We will use the majority of the results we obtain here in the development of our main subject, but some topics we discuss just out of pure interest or to improve our intuition.

First, let us generalize the definition of the Laplace operator to $\mathbb{E}^p$, where a point[2] x is given by the ordered pair $(x_1, x_2, \ldots, x_p)$.

Definition The *Laplace operator* in $\mathbb{E}^p$ is given by

$$\Delta_p \overset{\text{def}}{=} \sum_{i=1}^{p} \frac{\partial^2}{\partial x_i^2} \, . \tag{2.1}$$

The *del operator* in $\mathbb{E}^p$ is the vector operator

$$\nabla_p \overset{\text{def}}{=} \left(\frac{\partial}{\partial x_1}, \frac{\partial}{\partial x_2}, \ldots, \frac{\partial}{\partial x_p} \right) \, .$$

---

[1] It is often the case that by $\mathbb{R}^p$ many authors mean $\mathbb{E}^p$.
[2] We will not place vector arrows above points x in $\mathbb{R}^p$ or $\mathbb{E}^p$.

2.1 Rotations in $\mathbb{E}^p$

Let us quickly consider orthogonal rotations of the coordinate axes in $\mathbb{E}^p$. Such rotations leave the lengths of vectors unchanged. Indeed, the length of a vector is a geometric quantity; rotating the coordinate system we use to describe the vector leaves its length invariant. In fact, in a more abstract setting, we could define a rotation to be any transformation of coordinates that leaves the lengths of vectors unchanged.

In what follows, we let x denote a column vector[3] $(x_1, x_2, \ldots, x_p)^t$ in $\mathbb{E}^p$ and use $\langle \cdot, \cdot \rangle$ to represent the dot product of two vectors. The fact that a rotation matrix R leaves the length of x invariant means $\langle Rx, Rx \rangle = \langle x, x \rangle$. Moreover, since the dot product between any two vectors x, y can be written as

$$\langle x, y \rangle = \frac{1}{2} \big(\langle x + y, x + y \rangle - \langle x, x \rangle - \langle y, y \rangle \big),$$

it follows that coordinate rotations leave all dot products invariant.

Notice further that we can write dot products such as $\langle x, y \rangle$ as matrix products $y^t x$. In this notation, the requirement that $\langle Rx, Ry \rangle = \langle x, y \rangle$ translates into the necessity of $(Ry)^t (Rx) = y^t x$, or $y^t R^t R x = y^t x$. Since this equation must hold for all $x, y \in \mathbb{E}^p$, we can conclude that any rotation matrix R must satisfy $R^t R = I$, where I is the identity matrix. A matrix R that satisfies this requirement is called *orthogonal*.

Now we can verify our claim in the first few sentences of this booklet that the Laplace operator Δ_p remains unchanged after being subjected to a rotation of coordinates.

Proposition 2.1 *The Laplace operator Δ_p is invariant under coordinate rotations. That is, if R is a rotation matrix and $x' = Rx$, then $\Delta_p' = \Delta_p$, i.e.*

$$\sum_{j=1}^{p} \left(\frac{\partial}{\partial x_j'} \right)^2 = \sum_{j=1}^{p} \left(\frac{\partial}{\partial x_j} \right)^2.$$

---

[3]The superscript t denotes the operation of matrix transposition.

Proof This can be proved very easily by noticing that $\Delta_p = \nabla_p \cdot \nabla_p$ is a dot product of vector operators. Since all dot products are unchanged by coordinate rotations, we can conclude that Δ_p is not affected by any rotation R.

In case the reader is not satisfied with this quick justification, let us compute Δ_p', the Laplace operator after application of a rotation of coordinates R. Since R is a rotation matrix, it is orthogonal, i.e. $RR^t = I$. Then, using the chain rule,

$$
\Delta_p = \sum_{j=1}^{p} \left(\frac{\partial}{\partial x_j} \right)^2 = \sum_{j=1}^{p} \left[\left(\sum_{k=1}^{p} \frac{\partial x_k'}{\partial x_j} \frac{\partial}{\partial x_k'} \right) \left(\sum_{\ell=1}^{p} \frac{\partial x_\ell'}{\partial x_j} \frac{\partial}{\partial x_\ell'} \right) \right]
$$
$$
= \sum_{j=1}^{p} \left[\left(\sum_{k=1}^{p} R_{kj} \frac{\partial}{\partial x_k'} \right) \left(\sum_{\ell=1}^{p} R_{\ell j} \frac{\partial}{\partial x_\ell'} \right) \right],
$$

so

$$
\Delta_p = \sum_{k,\ell=1}^{p} \frac{\partial}{\partial x_k'} \frac{\partial}{\partial x_\ell'} \left(\sum_{j=1}^{p} R_{kj} R_{j\ell}^t \right) = \sum_{k=1}^{p} \left(\frac{\partial}{\partial x_k'} \right)^2 = \Delta_p'.
$$

This proves the proposition.   ■

2.2 Spherical Coordinates in p Dimensions

Now, in order to develop some experience and intuition working in higher-dimensional spaces, we will develop the spherical coordinate system for $\mathbb{E}^p$ in considerable detail. In particular, we will use an inductive technique to come up with the expression of the spherical coordinates in p dimensions in terms of the corresponding Cartesian coordinates.

We will let our space have axes denoted $x_1, x_2, \ldots$ First, in two dimensions, spherical coordinates are just the polar coordinates given in (1.5),

$$
r = \sqrt{x_1^2 + x_2^2}, \qquad r \in [0, \infty),
$$
$$
\phi = \tan^{-1} \frac{x_2}{x_1}, \qquad \phi \in [0, 2\pi),
$$

where r is the distance from the origin and ϕ is the azimuthal angle[4] in the plane that measures the rotation around the origin. The inverse transformation is

$$x_1 = r \cos\phi, \quad x_2 = r \sin\phi.$$

When we move to three dimensions we add an axis, naming it x_3, perpendicular to the plane. Now, the polar coordinates above can only define a location in the plane; thus, they only tell us on which vertical line (i.e., line parallel to the x_3-axis) we lie, as we can see in Figure 2.1 with $p = 3$. To pinpoint our location on this line, we introduce a new angle θ_1. When we also redefine r to be the three-dimensional distance from the origin, we have the spherical coordinates given in (1.10),

$$r = \sqrt{x_1^2 + x_2^2 + x_3^2}, \qquad r \in [0, \infty),$$
$$\phi = \tan^{-1}\frac{x_2}{x_1}, \qquad \phi \in [0, 2\pi),$$
$$\theta_1 = \tan^{-1}\frac{\sqrt{x_1^2 + x_2^2}}{x_3}, \qquad \theta_1 \in \left[-\frac{\pi}{2}, \frac{\pi}{2}\right].$$

Now let us imagine moving to four dimensions by adding an axis — name it the x_4-axis — perpendicular to the three-dimensional space just discussed. The three-dimensional spherical coordinates given above can only define a location in $\mathbb{E}^3$, so they only tell us on which "vertical" line (i.e., line parallel to the x_4-axis) we lie, as in Figure 2.1 with $p = 4$. We thus introduce a new angle θ_2 to determine the location on this line. We redefine r to be the four-dimensional distance from the origin, and this completes the construction of the

---

[4]As written, this widely used expression of ϕ in terms of x_1, x_2 is incorrect. A more precise formula would use a two-argument $\tan^{-1}$ function to produce angles on the entire unit circle. However, for this and one or two similar formulas, we allow ourselves to be sloppy in this section for simplicity.

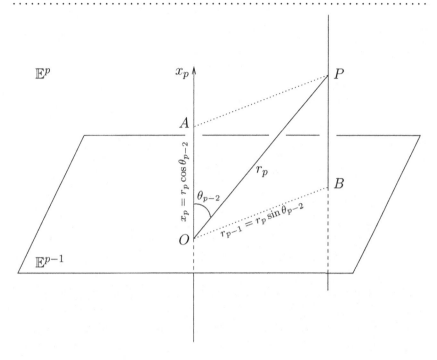

Figure 2.1: In going from $\mathbb{E}^{p-1}$ to $\mathbb{E}^p$, we visualize $\mathbb{E}^{p-1}$ as a plane and add a new perpendicular direction. We introduce a new angular coordinate θ_{p-2} to determine the location in the new direction.

four-dimensional spherical coordinates,

$$r = \sqrt{x_1^2 + x_2^2 + x_3^2 + x_4^2}, \qquad r \in [0, \infty),$$

$$\phi = \tan^{-1}\frac{x_2}{x_1}, \qquad \phi \in [0, 2\pi),$$

$$\theta_1 = \tan^{-1}\frac{\sqrt{x_1^2 + x_2^2}}{x_3}, \qquad \theta_1 \in \left[-\frac{\pi}{2}, \frac{\pi}{2}\right],$$

$$\theta_2 = \tan^{-1}\frac{\sqrt{x_1^2 + x_2^2 + x_3^2}}{x_4}, \qquad \theta_2 \in \left[-\frac{\pi}{2}, \frac{\pi}{2}\right],$$

where it should be clear from the figure that the new angle θ_2 only ranges from $-\pi/2$ to $\pi/2$.

In going from three to four dimensions, we mimicked the way in which we transitioned from two to three dimensions. We follow the same procedure each time we move up in dimension. For arbitrary p, this yields

$$r = \sqrt{x_1^2 + x_2^2 + \cdots + x_p^2},$$

$$\phi = \tan^{-1} \frac{x_2}{x_1},$$

$$\theta_1 = \tan^{-1} \frac{\sqrt{x_1^2 + x_2^2}}{x_3},$$

$$\vdots$$

$$\theta_{p-2} = \tan^{-1} \frac{\sqrt{x_1^2 + x_2^2 + \cdots + x_{p-1}^2}}{x_p},$$

where the ranges on the coordinates are as expected from the previous cases.

We have thus defined the spherical coordinates in p-dimensions in terms of the corresponding Cartesian coordinates. To write down the inverse relations we use projections, with Figure 2.1 as an aid. As before, we will derive these relations in detail for a few instructive cases before writing down the most general expressions.

We have already written down x_1, x_2 in terms of r, ϕ so let's use $\mathbb{E}^3$ as our first example. We imagine $\mathbb{E}^3$ as the direct sum of a two-dimensional plane $\mathbb{E}^2$ with the real line $\mathbb{E}$. Then, given the point P in $\mathbb{E}^3$ with spherical coordinates (r_3, ϕ, θ_1), the vector $\overrightarrow{OP}$ can be written as a sum $\overrightarrow{OA} + \overrightarrow{OB}$, where $\overrightarrow{OA}$ lies along the x_3-axis and has magnitude $r_3 \cos\theta_1$ and $\overrightarrow{OB}$ lies in the plane and has magnitude $r_2 = r_3 \sin\theta_1$. The point B thus has spherical coordinates (r_2, ϕ) in the plane, implying

$$x_1 = r_2 \cos\phi \quad \text{and} \quad x_2 = r_2 \sin\phi,$$

so

$$
\begin{align}
x_1 &= r_3 \sin\theta_1 \cos\phi, & (2.2) \\
x_2 &= r_3 \sin\theta_1 \sin\phi, & (2.3) \\
x_3 &= r_3 \cos\theta_1. & (2.4)
\end{align}
$$

Let us examine the situation in $\mathbb{E}^4$ using the same technique. Given the point P with radial distance r_4 from the origin, we decompose $\overrightarrow{OP}$ into two vectors $\overrightarrow{OA} + \overrightarrow{OB}$, where $\overrightarrow{OA}$ lies along the x_4-axis and has magnitude $r_4 \cos\theta_2$ and $\overrightarrow{OB}$ lies in the "plane" and has magnitude $r_3 = r_4 \sin\theta_2$. The point B thus has spherical coordinates (r_3, ϕ, θ_1) in the three-dimensional space, so x_1, x_2, x_3 are as in (2.2)–(2.4). Therefore

$$
\begin{aligned}
x_1 &= r_4 \sin\theta_2 \sin\theta_1 \cos\phi, \\
x_2 &= r_4 \sin\theta_2 \sin\theta_1 \sin\phi, \\
x_3 &= r_4 \sin\theta_2 \cos\theta_1, \\
x_4 &= r_4 \cos\theta_2.
\end{aligned}
$$

If the reader has understood the previous constructions, the expressions in p dimensions should be evident. Given the radius r_p in $\mathbb{E}^p$, we project the position vector onto $\mathbb{E}^{p-1}$, obtaining the radius in the subspace $\mathbb{E}^{p-1}$ given by $r_{p-1} = r_p \sin\theta_{p-2}$. In this way we can perform a series of projections to get down to a space in which we already know the relations. This procedure leads to (dropping the subscript on r_p)

$$
\begin{aligned}
x_1 &= r \sin\theta_{p-2} \sin\theta_{p-3} \cdots \sin\theta_3 \sin\theta_2 \sin\theta_1 \cos\phi, \\
x_2 &= r \sin\theta_{p-2} \sin\theta_{p-3} \cdots \sin\theta_3 \sin\theta_2 \sin\theta_1 \sin\phi, \\
x_3 &= r \sin\theta_{p-2} \sin\theta_{p-3} \cdots \sin\theta_3 \sin\theta_2 \cos\theta_1, \\
x_4 &= r \sin\theta_{p-2} \sin\theta_{p-3} \cdots \sin\theta_3 \cos\theta_2, \\
&\ \ \vdots \\
x_{p-1} &= r \sin\theta_{p-2} \cos\theta_{p-3}, \\
x_p &= r \cos\theta_{p-2}.
\end{aligned}
$$

With this chapter as an exception, we will rarely refer explicitly to the angles $\phi, \theta_1, \ldots, \theta_{p-2}$. However, we will frequently use $r = \sqrt{x_1^2 + \cdots + x_p^2}$.

2.3   The Sphere in Higher Dimensions

We will give definitions of the sphere and the ball in an arbitrary number of dimensions that are analogous to the definitions of the familiar sphere and ball we visualize embedded in $\mathbb{E}^3$.

Definition The $(p-1)$-*sphere of radius δ centered at x_0* is the set

$$S_\delta^{p-1}(x_0) \overset{\text{def}}{=} \{x \in \mathbb{E}^p : |x - x_0| = \delta\}.$$

The *unit $(p-1)$-sphere*[5] is the set $S^{p-1} \overset{\text{def}}{=} S_1^{p-1}(0)$.

Definition The *open p-ball of radius δ centered at x_0* is the set

$$B_\delta^p(x_0) \overset{\text{def}}{=} \{x \in \mathbb{E}^p : |x - x_0| < \delta\}.$$

The *open unit p-ball* is the set $B^p \overset{\text{def}}{=} B_1^p(0)$. The *closed p-ball* is the set $\bar{B}^p \overset{\text{def}}{=} B^p \cup S^{p-1}$.

Notice we call the sphere that we picture embedded in $\mathbb{E}^p$ the $(p-1)$-sphere because it is $(p-1)$-dimensional. It requires $p-1$ angles to define one's location on the sphere, as we saw in Section 2.2. However, it requires p coordinates to locate a point in the ball because we must also specify the radial distance r from the center; this justifies the notation B^p. Notice also that the $(p-1)$-sphere is the boundary of the p-ball when we think of these as subsets of $\mathbb{E}^p$; we write $S^{p-1} = \partial B^p$.

Let us compute the surface area of the $(p-1)$-sphere. Towards this goal, we recall that that the *gamma function* is defined by

$$\Gamma(z) = \int_0^\infty e^{-t} t^{z-1}\, dt, \tag{2.5}$$

for any $z \in \mathbb{C}$ such that $\mathrm{Re}(z) > 0$. Then

---

[5]Frequently, we will drop the "unit," though it is still implied.

Lemma 2.2 *For all $p \in \mathbb{C}$ such that $Re(p) > 0$, we have*

$$\int_0^{+\infty} e^{-r^2} r^{p-1} \, dr = \frac{1}{2} \Gamma\left(\frac{p}{2}\right).$$

Proof Using the substitution $u = r^2$,

$$\int_0^{+\infty} e^{-r^2} r^{p-1} \, dr = \int_0^{\infty} e^{-u} u^{\frac{p-1}{2}} \frac{dr}{2\sqrt{u}}$$

$$= \frac{1}{2} \int_0^{+\infty} e^{-u} u^{\frac{p}{2}-1} \, du = \frac{1}{2} \Gamma\left(\frac{p}{2}\right),$$

as sought.   ∎

Proposition 2.3 *If Ω_{p-1} denotes the solid angle in $\mathbb{E}^p$ (equivalent numerically to the surface area) of S^{p-1}), then*

$$\Omega_{p-1} = \frac{2\pi^{p/2}}{\Gamma(p/2)}.$$

Before proving this we will digress slightly and discuss how the surface area of the $(p-1)$-sphere relates to the volume of the p-ball. First, consider the p-ball of radius r centered at the origin. Notice that the radius r completely determines such a ball. If we were to determine the volume V_p of this p-ball, we would obtain $V_p = c r^p$ where c is some constant. Here, we have determined that $V_p \propto r^p$ because V_p must have dimensions of [length]^p and the only variable characterizing the p-ball is r (which has dimensions of length). Now if we differentiate V_p with respect to r, we get

$$\frac{dV_p}{dr} = (p-1) \, c \, r^{p-1}.$$

Thus, if the radius of a p-ball changes by an infinitesimal amount δr, its volume will change by some infinitesimal amount δV_p, and

$$\delta V_p = (p-1)\, c\, r^{p-1}\, \delta r. \tag{2.6}$$

But in this case, the small change in volume δV should equal the surface area $A_{p-1}(r)$ of the p-ball multiplied by the small change in radius δr,

$$\delta V_p = A_{p-1}(r)\, \delta r. \tag{2.7}$$

From (2.6) and (2.7) we see that

$$A_{p-1}(r) = (p-1)\, c\, r^{p-1}\,,$$

and if we let $\Omega_{p-1} = (p-1)\, c$ denote the numerical value of the surface area when $r = 1$ (as in the statement of Proposition 2.3), we get[6]

$$A_{p-1}(r) \;=\; \Omega_{p-1}\, r^{p-1}\,. \tag{2.8}$$

We see that if we want to carry out an integral over $\mathbb{E}^p$ when the integrand depends only on r, we can use the differential volume element

$$dV_p \;=\; A_{p-1}(r)\, dr = r^{p-1}\, \Omega_{p-1}\, dr\,.$$

Now we are ready to prove Proposition 2.3.

Proof Consider the integral,

$$J \;=\; \int\limits_{-\infty}^{\infty} dx_1 \int\limits_{-\infty}^{\infty} dx_2 \cdots \int\limits_{-\infty}^{\infty} dx_p\, e^{-(x_1^2 + x_2^2 + \cdots + x_p^2)}\,.$$

This is really

$$J \;=\; \left(\int\limits_{-\infty}^{\infty} e^{-x^2}\, dx \right)^{p} = \left(\sqrt{\pi} \right)^{p}\,.$$

---

[6]We should stress that, since a physicist assigns physical dimensions to each and every quantity, s/he would differentiate between the solid angle Ω_{p-1} and the surface area A_{p-1} of the $(p-1)$-sphere. Although numerically the two quantities are equal for the unit sphere since $r = 1$, they are different quantities since Ω_{p-1} is dimensionless while A_{p-1} has dimensions of [length]$^{p-1}$.

Using spherical coordinates however, we can write

$$J \;=\; \int\limits_{S^{p-1}} \int\limits_0^{+\infty} dV_p \, e^{-r^2} \;=\; \Omega_{p-1} \int\limits_0^{\infty} e^{-r^2} \, r^{p-1} \, dr \,,$$

since Ω_{p-1} is just a constant. Therefore

$$\Omega_{p-1} \;=\; \frac{\pi^{p/2}}{\int_0^{\infty} e^{-r^2} \cdot r^{p-1} \, dr} \,,$$

and, with the help of Lemma 2.2 we arrive at the advertised result. ∎

Remark As a check, we can use this formula to determine the surface area (circumference) of the 1-sphere (circle) as well as the surface area of the 2-sphere, which is the familiar sphere that we embed in $\mathbb{E}^3$. As expected,

$$\Omega_1 \;=\; 2\pi, \quad \Omega_2 \;=\; \frac{2\pi^{3/2}}{\Gamma(3/2)} \;=\; 4\pi \,,$$

using $\Gamma(3/2) = \sqrt{\pi}/2$. It is interesting to consider the 0-sphere, i.e., the sphere that we visualize embedded in $\mathbb{E}$. S^0 consists of all points on the real line that are unit distance from the origin, so $S^0 = \{-1, 1\}$. Using Lemma 2.3, we find $\Omega_0 = 2$, since $\Gamma(1/2) = \sqrt{\pi}$. That is, the surface area of just two points on the real line is finite! This is consistent with standard calculus, though. On the real line, the radial distance is the absolute value of x. Hence, the concept of a function $f(r)$ possessing spherical symmetry coincides with the concept of an even function, $f(x) = f(-x)$. Then

$$\int\limits_{-\infty}^{+\infty} f(x) \, dx \;=\; 2 \int\limits_0^{\infty} f(x) \, dx \;=\; \int\limits_0^{\infty} f(r) \, \Omega_0 \, r^0 \, dr \,.$$

2.4   Arc Length in Spherical Coordinates

Now, let us compute the formula for the differential arc length in $\mathbb{E}^p$. That is, given the points P and P' with coordinates $(r, \phi, \theta_1, \ldots, \theta_{p-2})$ and $(r + dr, \phi + d\phi, \theta_1 + d\theta_1, \ldots, \theta_{p-2} + d\theta_{p-2})$ respectively, we derive a formula for the distance $d\ell_p$ between them.

In this section, if you prefer, use the physicists' interpretation of the differentials: infinitesimals. That is, interpret any differential as a tiny quantity — in fact, tiny enough so that the errors involved in the approximations (such as assuming the small portion of a curve between P and P' is almost straight) are as small as desired (say, less than some given $\epsilon > 0$).

Knowing the spherical coordinates of the point P, we can write its Cartesian coordinates as

$$x_i = r\, \xi_i, \quad i = 1, 2, \ldots, p,$$

where ξ is a unit vector in $\mathbb{E}^p$. Also, knowing the difference between the spherical coordinates of the points P and P' implies knowing the difference between the Cartesian coordinates of the two points:

$$dx_i = dr\, \xi_i + r\, d\xi_i, \quad i = 1, 2, \ldots, p, \qquad (2.9)$$

where the $d\xi_i$ can be calculated using the Leibnitz rule for the differentiation of a product. However, there is no need to compute the $d\xi_i$ explicitly. Since ξ is a unit vector, by differentiating $\xi^2 = 1$, we conclude that the vector $d\xi$ is perpendicular to ξ. That is,

$$\sum_{i=1}^{p} \xi_i\, d\xi_i = 0.$$

By squaring expression (2.9) and taking the sum over i, we find

$$\sum_{i=1}^{p} dx_i^2 = dr^2 \sum_{i=1}^{p} (\xi_i)^2 + r^2 \sum_{i=1}^{p} d\xi_i^2 + 2\, r\, dr \sum_{i=1}^{p} \xi_i\, d\xi_i.$$

The left-hand side is the square length $d\ell_p^2$. In the right-hand side, the sum in the third term is zero as explained above. Also, the sum in the first term of the right-hand side is the square length of ξ — that is, 1. The remaining sum is the square length of $d\xi$ (scaled by a factor r^2) which lies completely inside the $(p-1)$-sphere. Let's indicate this length by ds_{p-1}. Thus we have derived the *Pythagorean Theorem* for p dimensions:

$$d\ell_p^2 \;=\; dr^2 + r^2\, ds_{p-1}^2\,.$$

We move to compute the length ds_{p-1}. Let S^{p-1} be the $(p-1)$-sphere. Any point on this sphere is represented by a vector ξ with components:

$$
\begin{aligned}
\xi_1 &= \sin\theta_{p-2}\,\sin\theta_{p-3}\cdots\sin\theta_3\,\sin\theta_2\,\sin\theta_1\,\cos\phi\,,\\
\xi_2 &= \sin\theta_{p-2}\,\sin\theta_{p-3}\cdots\sin\theta_3\,\sin\theta_2\,\sin\theta_1\,\sin\phi\,,\\
\xi_3 &= \sin\theta_{p-2}\,\sin\theta_{p-3}\cdots\sin\theta_3\,\sin\theta_2\,\cos\theta_1\,,\\
\xi_4 &= \sin\theta_{p-2}\,\sin\theta_{p-3}\cdots\sin\theta_3\,\cos\theta_2\,,\\
&\;\;\vdots\\
\xi_{p-1} &= \sin\theta_{p-2}\,\cos\theta_{p-3}\,,\\
\xi_p &= \cos\theta_{p-2}\,.
\end{aligned}
$$

Let also S^{p-2} be a $(p-2)$-sphere. Any point on this sphere is represented by a vector η with components:

$$
\begin{aligned}
\eta_1 &= \sin\theta_{p-3}\cdots\sin\theta_3\,\sin\theta_2\,\sin\theta_1\,\cos\phi\,,\\
\eta_2 &= \sin\theta_{p-3}\cdots\sin\theta_3\,\sin\theta_2\,\sin\theta_1\,\sin\phi\,,\\
\eta_3 &= \sin\theta_{p-3}\cdots\sin\theta_3\,\sin\theta_2\,\cos\theta_1\,,\\
\eta_4 &= \sin\theta_{p-3}\cdots\sin\theta_3\,\cos\theta_2\,,\\
&\;\;\vdots\\
\eta_{p-1} &= \cos\theta_{p-3}\,.
\end{aligned}
$$

It is thus straightforward to notice the relation:

$$
\begin{aligned}
\xi_i &= (\sin\theta_{p-2})\,\eta_i\,, \quad i = 1, 2, \ldots, p-1\,,\\
\xi_p &= \cos\theta_{p-2}\,.
\end{aligned}
$$

Notice that the subset of points of S^{p-1} having $\theta_{p-2} = \pi/2$ (we call it the equator of S^{p-1}) is S^{p-2}. Differentiating the previous equations, squaring them and, finally, adding them we find:

$$\sum_{i=1}^{p} d\xi_i^2 = d\theta_{p-2}^2 + \sin^2 \theta_{p-2} \sum_{i=1}^{p-1} d\eta_i^2 \,,$$

or

$$ds_{p-1}^2 = d\theta_{p-2}^2 + \sin^2 \theta_{p-2} \, ds_{p-2}^2 \,.$$

Using this equation inductively, we conclude that

$$\begin{aligned}
ds_{p-1}^2 = \; & d\theta_{p-2}^2 + \sin^2 \theta_{p-2} \, d\theta_{p-3}^2 + \sin^2 \theta_{p-2} \sin^2 \theta_{p-3} \, d\theta_{p-4}^2 + \cdots \\
& + \sin^2 \theta_{p-2} \sin^2 \theta_{p-3} \cdots \sin^2 \theta_2 \, d\theta_1^2 \\
& + \sin^2 \theta_{p-2} \sin^2 \theta_{p-3} \cdots \sin^2 \theta_1 \, d\phi^2 \,.
\end{aligned}$$

$$(2.10)$$

2.5 The Divergence Theorem in $\mathbb{E}^p$

Let us now return to the use of Cartesian coordinates to give an intuitive "derivation" of the divergence theorem in p dimensions, which we will use numerous times in Chapter 4. For brevity, our justification of this theorem will be rather physical. For a more mathematical treatment, the interested reader should consult a calculus book (for example, [18]) for the theorem in $\mathbb{E}^3$ and see if s/he can generalize the proof there to $\mathbb{E}^p$. For a rigorous proof of the divergence theorem in p dimensions, she/he may consult an analysis text (for example, [16]).

Theorem 2.4 *Let $\vec{F}$ be a continuously differentiable vector field defined in the neighborhood of some closed, bounded domain V in $\mathbb{E}^p$ which has smooth boundary ∂V. Then*[7]

$$\int_V \nabla \cdot \vec{F} \, d^p x = \oint_{\partial V} \vec{F} \cdot \hat{n} \, d\sigma \,,$$

$$(2.11)$$

where $\hat{n}$ is the unit outward normal vector on ∂V and $d\sigma$ is the differential element of surface area on ∂V.

---
[7]Here and from now on, $d^p x = dx_1 dx_2 \cdots dx_p$.

"Proof" We will interpret $\vec{F}$ as the flux density of some p-dimensional fluid moving through the volume V. In unit time at a point x, the volume of fluid which flows past an arbitrarily oriented unit surface with unit normal vector $\hat{n}$ is given by $\vec{F}(x) \cdot \hat{n}$.

Let us fill the interior of V with a "grid" of disjoint boxes, none intersecting the boundary ∂V — see Figure 2.2. If the boxes are comparable in size to V, they will not make a complete covering of the interior, since many regions of V near ∂V will remain uncovered; however, as the lengths of the box edges approach zero, the entire interior of V can be covered by these boxes. We will consider one small

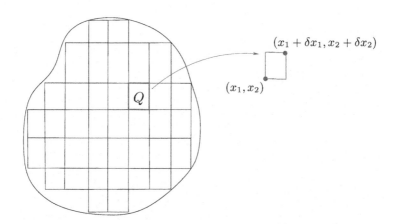

Figure 2.2: An example of a domain V in $\mathbb{E}^2$. A similar picture would describe the situation in arbitrary $\mathbb{E}^p$. We consider one box Q with bottom-left corner at (x_1, x_2) and top-right corner at $(x_1 + \delta x_1, x_2 + \delta x_2)$.

box Q in the interior of V, say with one corner at $(x_1, x_2, \ldots, x_p)$ and another at $(x_1 + \delta x_1, x_2 + \delta x_2, \ldots, x_p + \delta x_p)$, where the δx_i are all positive. Consider the integral

$$J = \oint_{\partial Q} \vec{F} \cdot \hat{n} \, d\sigma \,.$$

The boundary ∂Q consists of $2p$ planes, two orthogonal to each coordinate axis. The two planes orthogonal to the x_1 axis have surface area equal to $\delta x_2 \, \delta x_3 \cdots \delta x_p$. Since $\vec{F}$ is continuous, the integral of $\vec{F} \cdot \hat{n}$ over these surfaces can be written as[8]

$$\vec{F}(x_1 + \delta x_1, x_2, \ldots, x_p) \cdot \hat{x}_1 \, \delta x_2 \, \delta x_3 \cdots \delta x_p \,,$$

and

$$\vec{F}(x_1, x_2, \ldots, x_p) \cdot (-\hat{x}_1) \, \delta x_2 \, \delta x_3 \cdots \delta x_p \,,$$

since $\vec{F}$ changes very little over this plane as $\delta x_1, \delta x_2, \ldots, \delta x_p \to 0$. Now, there was nothing special about the x_1-axis, so we see that for each x_i-axis, the integral J will include a term

$$\left[\vec{F}(x_1, \ldots, x_i + \delta x_i, \ldots, x_p) - \vec{F}(x_1, \ldots, x_i, \ldots, x_p) \right] \cdot \hat{x}_i \, \frac{\delta x_1 \cdots \delta x_p}{\delta x_i} \,.$$

This can be rewritten as

$$\left[\frac{\vec{F}(x_1, \ldots, x_i + \delta x_i, \ldots, x_p) - \vec{F}(x_1, \ldots, x_i, \ldots, x_p)}{\delta x_i} \right] \cdot \hat{x}_i \, \delta x_1 \cdots \delta x_p \,,$$

which becomes

$$\frac{\partial \vec{F}(x_1, \ldots, x_p)}{\partial x_i} \cdot \hat{x}_i \, \delta x_1 \cdots \delta x_p \quad \text{or} \quad \frac{\partial F_i(x_1, \ldots, x_p)}{\partial x_i} \, \delta x_1 \cdots \delta x_p$$

in the limit with which we are concerned. Putting all these terms together, we get

$$J = \sum_{i=1}^{p} \frac{\partial F_i(x_1, \ldots, x_p)}{\partial x_i} \, \delta x_1 \cdots \delta x_p \,.$$

But this is just

$$\nabla_p \cdot \vec{F}(x_1, \ldots, x_p) \, \delta x_1 \cdots \delta x_p = \int_Q \nabla \cdot \vec{F} \, d^p x$$

---

[8]Here, $\hat{x}_1$ is a unit vector in the direction of increasing x_1.

in our limit. We have thus shown that

$$\int_Q \nabla \cdot \vec{F} \, d^p x = \oint_{\partial Q} \vec{F} \cdot \hat{n} \, d\sigma$$

for the special case of an infinitesimal box Q inside V. Let us now sum this result over all the boxes inside V,

$$\sum_{Q_i} \int_{Q_i} \nabla \cdot \vec{F} \, d^p x = \sum_{Q_i} \oint_{\partial Q_i} \vec{F} \cdot \hat{n} \, d\sigma , \qquad (2.12)$$

where we are concerned with the limit as the lengths of the box edges go to zero and the number of boxes in the covering approaches infinity. In this limit, the volume of the uncovered regions of V near ∂V approaches zero. Since $\vec{F}$ is continuously differentiable, it follows that $\nabla \cdot \vec{F}$ is continuous and thus bounded over the closed and bounded region V. Thus, the integral of $\nabla \cdot \vec{F}$ over regions in V not covered by boxes approaches zero. The left side of (2.12) therefore becomes

$$\sum_{Q_i} \int_{Q_i} \nabla \cdot \vec{F} \, d^p x \longrightarrow \int_V \nabla \cdot \vec{F} \, d^p x \qquad (2.13)$$

in this limit. Now, in the right side of (2.12), let us consider all the planes which bound the boxes Q_i that are included in the sum. Each "interior" plane appears twice in the sum, once as the "right" side of one box and a second time as the "left" side of another box. In each of these appearances, both $\vec{F}$ and $d\sigma$ remain the same but the vector $\hat{n}$ shows up with opposite sign. Thus, the integral over all the interior planes vanishes. The only terms that are not canceled in the integral on the right side of (2.12) make up the integral over the "exterior" planes, which we write as $\partial \bigcup Q_i$. That is,

$$\sum_{Q_i} \oint_{\partial Q_i} \vec{F} \cdot \hat{n} \, d\sigma = \oint_{\partial(\cup Q_i)} \vec{F} \cdot \hat{n} \, d\sigma ,$$

so using (2.13),

$$\int_V \nabla \cdot \vec{F} \, d^p x = \oint_{\partial(\cup Q_i)} \vec{F} \cdot \hat{n} \, d\sigma . \qquad (2.14)$$

Now we are very close to (2.11), but there is one difficulty due to the fact that the outward normal vector $\hat{n}$ for $\partial \bigcup_i Q_i$ always points along one of the coordinate axes while $\hat{n}$ can point in any direction for ∂V. To reconcile this difference, we realize that integrating $\vec{F} \cdot \hat{n}$ over a closed surface gives us the volume of fluid that has passed through this surface in unit time. The same volume of fluid must pass through both $\partial \bigcup_i Q_i$ and ∂V since $\partial \bigcup_i Q_i \to \partial V$. Therefore,

$$\oint_{\partial(\cup_i Q_i)} \vec{F} \cdot \hat{n} \, d\sigma \;=\; \oint_{\partial V} \vec{F} \cdot \hat{n} \, d\sigma \, .$$

Combining this with (2.14) completes the proof.   ∎

2.6 Δ_p in Spherical Coordinates

To compute Δ_p in spherical coordinates, we could use the chain rule. This would involve converting all the derivatives with respect to x_i in (2.1) to derivatives with respect to spherical coordinates as we did for Δ_2 and Δ_3 in Section 1.1. Such an undertaking would be quite messy, however, and adds little additional insight.

Another approach would be to use the general formula from differential geometry[9]

$$\Delta \;=\; \frac{1}{\sqrt{g}} \, \partial_\mu \sqrt{g} g^{\mu\nu} \partial_\nu \, , \tag{2.15}$$

where $g_{\mu\nu}$ is the metric tensor of the space of interest, $g = \det[g_{\mu\nu}]$ and $g^{\mu\nu}$ the inverse of the metric tensor. Since we have avoided this route thus far, we will not use it here either.

We can actually use an integration trick to determine the form of the p-dimensional Laplace operator Δ_p in spherical coordinates. This is the approach followed below.

Consider the p-dimensional cone-like region depicted in Figure 2.3. Let $\mathcal{D}$ be the shaded region (a truncated cone), which has left and right boundaries given by portions of $S_r^{p-1}(0)$ and $S_{r+\delta r}^{p-1}(0)$

---

[9]In this form, Δ is known as the *Laplace-Beltrami operator*.

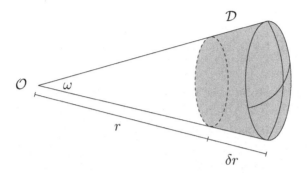

Figure 2.3: A region $\mathcal{D}$ in $\mathbb{E}^p$.

respectively. We are concerned with the limit where the quantity $\delta r \to 0$. This region spans a solid angle of ω, which means the surface area of the region is a fraction ω/Ω_{p-1} of the surface area of a complete sphere of the same radius. Using (2.8), this implies that the surface area of the left boundary is ωr^{p-1} while that of the right boundary is $\omega(r + \delta r)^{p-1}$.

Now let $f : \mathbb{E}^p \to \mathbb{E}$ be a twice continuously differentiable function, and use the divergence theorem in p dimensions to find[10]

$$\int_{\mathcal{D}} \Delta_p f \, d^p x = \int_{\mathcal{D}} \nabla_p \cdot (\nabla_p f) \, d^p x = \int_{\partial \mathcal{D}} \nabla_p f \cdot \hat{n} \, d\sigma = \int_{\partial \mathcal{D}} \frac{\partial f}{\partial n} \, d\sigma \,,$$

where $\hat{n}$ is the external unit normal vector to $\mathcal{D}$ and $d\sigma$ is a differential element of surface area. By the *mean value theorem for integration*,

$$\int_{\mathcal{D}} \Delta_p f \, d^p x = \Delta_p f(x^*) \int_{\mathcal{D}} d^p x = \Delta_p f(x^*) \cdot \mathrm{vol}(\mathcal{D}),$$

for some x^* in $\mathcal{D}$. But in the limit as the region of integration becomes infinitesimally small, it does not matter which x^* we use in $\mathcal{D}$ since f is continuous, so

$$\int_{\mathcal{D}} \Delta_p f \, d^p x \to \Delta_p f \cdot \mathrm{vol}(\mathcal{D}) \quad \text{as} \quad \mathrm{vol}(\mathcal{D}) \to 0 \,.$$

---

[10]We introduce here our notation for a normal derivative: $\frac{\partial f}{\partial n} = \nabla_p f \cdot \hat{n}$.

Thus

$$\Delta_p f = \lim_{\text{vol}(\mathcal{D}) \to 0} \left[\frac{1}{\text{vol}(\mathcal{D})} \int_{\partial \mathcal{D}} \left(\frac{\partial f}{\partial n} \right) d\sigma \right] .$$

Note that in this limit, the volume of $\mathcal{D}$ approaches $\omega r^{p-1} \delta r$. We can break up the integral in the numerator into one integral over the right bounding surface, one over the left bounding surface, and one over the lateral surface which we will denote by $\partial \mathcal{D}'$. Since in this limit the bounding surfaces are infinitesimally small, we can pull the integrands outside the integrals and rewrite the above equation as

$$\Delta_p f = \lim_{\substack{\delta r \to 0 \\ \omega \to 0}} \frac{\frac{\partial f}{\partial r}\big|_{r+\delta r} \omega (r + \delta r)^{p-1} - \frac{\partial f}{\partial r}\big|_r \omega r^{p-1}}{\omega r^{p-1} \delta r}$$

$$+ \lim_{\text{vol}(\mathcal{D}) \to 0} \left[\frac{\int_{\partial \mathcal{D}'} \left(\frac{\partial f}{\partial n} \right) d\sigma}{\text{vol}(\mathcal{D})} \right] ,$$

which becomes, using the binomial expansion,[11]

$$\lim_{\delta r \to 0} \frac{\frac{\partial f}{\partial r}\big|_{r+\delta r} \left[r^{p-1} + (p-1) r^{p-2} \delta r + \mathcal{O}(\delta r^2) \right] - \frac{\partial f}{\partial r}\big|_r r^{p-1}}{r^{p-1} \delta r} + C ,$$

or,

$$\Delta_p f = \lim_{\delta r \to 0} \left[\frac{\frac{\partial f}{\partial r}\big|_{r+\delta r} - \frac{\partial f}{\partial r}\big|_r}{\delta r} + \frac{p-1}{r} \frac{\partial f}{\partial r} \right] + C , \qquad (2.16)$$

where C stands for the contribution from the lateral surface term:

$$C = \lim_{\text{vol}(\mathcal{D}) \to 0} \left[\frac{1}{\text{vol}(\mathcal{D})} \int_{\partial \mathcal{D}'} \frac{\partial f}{\partial n} d\sigma \right] .$$

Notice that C contains directional derivatives only in directions orthogonal to the radial direction. Hence, this term thus contains no

---

[11] Here, $\mathcal{O}(\delta r^2)$ denotes terms which are such that $\frac{\text{terms}}{\delta r^2}$ remains finite as $\delta r \to 0$. In particular, $\frac{\text{terms}}{\delta r} \to 0$ as $\delta r \to 0$.

derivatives with respect to r and only derivatives with respect to the angles $\phi, \theta_1, \theta_2, \ldots, \theta_{p-2}$; we will indicate it by

$$\frac{1}{r^2} \Delta_{S^{p-1}} f \,,$$

where we determined the prefactor $1/r^2$ through dimensional analysis. We call the operator $\Delta_{S^{p-1}}$ the *spherical Laplace operator in* $p-1$ *dimensions*. Equation (2.16) thus implies the following proposition.

Proposition 2.5

$$\Delta_p \;=\; \frac{\partial^2}{\partial r^2} + \frac{p-1}{r}\frac{\partial}{\partial r} + \frac{1}{r^2}\Delta_{S^{p-1}} \,.$$

Alternatively, we may write the Laplace operator in p dimensions in the form

$$\Delta_p = \frac{1}{r^{p-1}}\frac{\partial}{\partial r}\left(r^{p-1}\frac{\partial}{\partial r}\right) + \frac{1}{r^2}\Delta_{S^{p-1}}. \tag{2.17}$$

Definition A function $f(x)$ is called *harmonic* if

$$\Delta_p f = 0\,. \tag{2.18}$$

Equation (2.18) is called the *Laplace equation*.

Using expression (2.17), it follows immediately that the only harmonic functions $f(r)$ depending only on the radial distance r in p dimensions are

$$f(r) = 1,\ r, \qquad\quad p = 1, \tag{2.19a}$$
$$f(r) = 1,\ \ln r, \qquad p = 2, \tag{2.19b}$$
$$f(r) = 1,\ r^{2-p}, \qquad p \geq 3. \tag{2.19c}$$

Notice that, for $p = 3$ we recover the well known gravitational and electrostatic potential $1/r$ (as we should).

Now that we are comfortable working in $\mathbb{E}^p$ let us move on to briefly study the subject of orthogonal polynomials which will be very useful when we come to our main topic of discussion.

2.7 Problems

1. Prove that if f is a harmonic function, then $g = r\,(\partial f/\partial r)$ is also harmonic.

2. For the p-dimensional ball of radius r, $x_1^2 + x_2^2 + \cdots + x_p^2 \le r^2$, write its volume V_p in the form

$$V_p = \int \cdots \int_{x^1 + x_2^2 + \cdots + x_p^2 \le r^2} dx_1\, dx_2 \cdots dx_p\,.$$

First write it in the form $V_p = r^n\, U_p$, and then show the recursion relation

$$U_p = U_{p-1} \int_{-\pi/2}^{\pi/2} \cos^n \theta\, d\theta\,.$$

From this prove that

$$V_p = r^p \frac{\pi^{p/2}}{\Gamma(p/2 + 1)}\,.$$

and show that it reproduces the expression for Ω_{p-1} given in this chapter.

3. From expression (2.10) extract the components of the metric tensor $[g_{\mu\nu}]$ for the $(p-1)$-dimensional sphere. Recall that, for any space, the metric tensor is defined from the differential length $d\ell$ between two points P and P' with coordinates[12] x^i and $x^i + dx^i$ where $i = 1, 2, \ldots, D$ and D is the dimensionality of the space:

$$d\ell^2 = \sum_{i,j=1}^{D} g_{\mu\nu}\, dx^i dx^j\,. \tag{2.20}$$

---

[12]At this point, consider the use of superscripts in the notation of the coordinates as an eccentricity. We will not be bothered with upper and lower indices in this book.

Let g stand for the determinant of the metric tensor, i.e. $g = \det[g_{ij}]$. Show that the surface area Ω_{p-1} of the $(p-1)$-dimensional sphere is given by the integral

$$\Omega_{p-1} \;=\; \int_0^{2\pi} d\phi \int_{-\pi/2}^{\pi/2} d\theta_1 \cdots \int_{-\pi/2}^{\pi/2} d\theta_{p-2} \,\sqrt{g}\,.$$

Compute the integral to verify the expression given for Ω_{p-1} in this chapter.

4. We can extend the idea of hyperspheres to hyperellipsoids. In particular, the $(p-1)$-dimensional ellipsoid or just the $(p-1)$-ellipsoid is defined by

$$\frac{x_1^2}{a_1^2} + \frac{x_2^2}{a_2^2} + \cdots + \frac{x_p^2}{a_p^2} \;=\; 1\,.$$

Obviously when $a_1 = a_2 = \cdots = a_p = R$ we recover a $(p-1)$-sphere with radius R. Find the volume bound by the ellipsoid.

5. Perhaps you are under the impression that as p increases the volume enclosed by the p-ball and its surface area increase. This is supported by the volumes $U_1 = 2$, $U_2 = \pi$, $U_3 = 4\pi/3$ and the areas $\Omega_0 = 2$, $\Omega_1 = 2\pi$, $\Omega_2 = 4\pi$. However, prove the counter-intuitive result

$$\lim_{p\to\infty} U_p \;=\; \lim_{p\to\infty} \Omega_p \;=\; 0\,.$$

It is obvious then that the sequences $\{U_p\}_{p\in\mathbb{N}}$ and $\{\Omega_p\}_{p\in\mathbb{N}}$ attain a maximum for some value(s) of p. Prove Nunemacher's results [14]:

(a) Among all unit balls B_p, the one with $p = 5$ has the greatest volume.

(b) Among all unit spheres S_p, the one with $p = 6$ has the greatest surface area.

6. In [19] Singmaster writes: "Some time ago, the following problem occurred to me: which fits better, a round peg in a square
hole or a square peg in a round hole? This can easily be solved
once one arrives at the following mathematical formulation of
the problem. Which is larger: the ratio of the area of a circle
to the area of the circumscribed square or the ratio of the area
of a square to the area of the circumscribed circle? One easily finds that the first ratio is $\pi/4$ and that the second is $2/\pi$.
Since the first is larger, we may conclude that a round peg fits
better in a square hole than a square peg fits in a round hole."

Prove Singmaster's theorem: The p-ball fits better in the p-
cube than the p-cube fits in the p-ball if and only if $p \leq 8$.

7. In $\mathbb{E}^3$, recall that

$$
\Delta_{S^2} = \frac{1}{\sin \theta} \frac{\partial}{\partial \theta} \left(\sin \theta \, \frac{\partial}{\partial \theta} \right) + \frac{1}{\sin^2 \theta} \frac{\partial^2}{\partial \phi^2} .
$$

Notice that $\frac{\partial^2}{\partial \phi^2}$ is the Laplacian of the 1-circle (think of it as
the equator of S^2 for $\theta = 0$). Therefore

$$
\Delta_{S^2} = \frac{1}{\sin \theta} \frac{\partial}{\partial \theta} \left(\sin \theta \, \frac{\partial}{\partial \theta} \right) + \frac{1}{\sin^2 \theta} \Delta_{S^1} .
$$

Generalize the previous relation to p dimensions. That is, consider the sphere S^{p-2} as the equator of S^{p-1} for $\theta_{p-2} = 0$. Then
prove that

$$
\Delta_{S^{p-1}} = \frac{1}{(\sin \theta_{p-2})^{p-2}} \frac{\partial}{\partial \theta_{p-2}} \left[(\sin \theta_{p-2})^{p-2} \frac{\partial}{\partial \theta_{p-2}} \right]
$$
$$
+ \frac{1}{(\sin \theta_{p-2})^2} \Delta_{S^{p-2}} , \quad (2.21)
$$

and thus

$$
\begin{aligned}
\Delta_{S^{p-1}} ={}& \frac{1}{(\sin\theta_{p-2})^{p-2}}\frac{\partial}{\partial\theta_{p-2}}\left[(\sin\theta_{p-2})^{p-2}\frac{\partial}{\partial\theta_{p-2}}\right]\\
&+\frac{1}{(\sin\theta_{p-2})^2}\frac{1}{(\sin\theta_{p-3})^{p-3}}\frac{\partial}{\partial\theta_{p-3}}\left[(\sin\theta_{p-3})^{p-3}\frac{\partial}{\partial\theta_{p-3}}\right]\\
&+\frac{1}{(\sin\theta_{p-2})^2}\frac{1}{(\sin\theta_{p-3})^2}\frac{1}{(\sin\theta_{p-4})^{p-4}}\frac{\partial}{\partial\theta_{p-4}}\left[(\sin\theta_{p-4})^{p-4}\frac{\partial}{\partial\theta_{p-4}}\right]\\
&+\dots\\
&+\frac{1}{(\sin\theta_{p-2})^2}\frac{1}{(\sin\theta_{p-3})^2}\cdots\frac{1}{(\sin\theta_2)^2}\frac{1}{(\sin\theta_1)^1}\frac{\partial}{\partial\theta_1}\left[(\sin\theta_1)^1\frac{\partial}{\partial\theta_{p-4}}\right]\\
&+\frac{1}{(\sin\theta_{p-2})^2}\frac{1}{(\sin\theta_{p-3})^2}\cdots\frac{1}{(\sin\theta_2)^2}\frac{1}{(\sin\theta_1)^2}\frac{\partial^2}{\partial\phi^2}.
\end{aligned}
$$

Chapter 3

Orthogonal Polynomials

This chapter provides the reader with several useful facts from the theory of orthogonal polynomials that we will require in order to prove some of the theorems in our main discussion. In what follows, we will assume the reader has taken an introductory course in linear algebra and, in particular, is familiar with inner product spaces and the Gram-Schmidt orthogonalization procedure. If this is not the case, we refer him or her to [8].

3.1   Orthogonality and Expansions

We will deal with a set $\mathcal{P}$ where the members are real polynomials of finite degrees in one variable defined on some interval[1] I. With vector addition defined as the usual addition of polynomials and scalar multiplication defined as ordinary multiplication by real numbers, it is easy to check that we have, so far, a well defined vector space

Let us put more structure on this space by adding an inner product. For any $p, q \in \mathcal{P}$, define a function $\langle \cdot, \cdot \rangle_w : \mathcal{P} \times \mathcal{P} \to \mathbb{R}$ by

$$\langle p, q \rangle_w \overset{\text{def}}{=} \int_I p(x)q(x)w(x)\,dx\,, \tag{3.1}$$

---

[1]For now, we allow I to be finite or infinite and open, closed, or neither.

where w, called a *weight function*, is some positive function defined on I such that

$$\int_I r(x)w(x)\,dx$$

exists and is finite for all polynomials $r \in \mathcal{P}$. Notice that, since the product of any two polynomials is itself a polynomial, the above requirement ensures the existence of $\langle p, q \rangle_w$ for all polynomials p, q. Moreover, the function (3.1) is symmetric and linear in both arguments and, since w is positive, $\langle r, r \rangle_w \geq 0$ for all polynomials r. Finally, since r^2 is continuous for all polynomials r,

$$\langle r, r \rangle_w = \int_I r(x)^2\,w(x)\,dx = 0\,,$$

if and only if $r(x) = 0$ for all $x \in I$. For suppose $r(x_0) \neq 0$ for some $x_0 \in I$. By continuity, r must not vanish in some neighborhood about x_0, and thus the above integral will acquire a positive value. Therefore, $\mathcal{P}$ together with the function (3.1) is a well defined inner product space. We call $\langle \cdot, \cdot \rangle_w$ the *inner product with respect to the weight w*.

The reader should recall the following two inequalities that hold in any inner product space. We state them here for reference without any proofs which may be found in [8].

Proposition 3.1 (Cauchy-Schwarz Inequality) *For any vectors p, q in an inner product space $\mathcal{P}$, we have $|\langle p, q \rangle| \leq \|p\| \|q\|$.*

Proposition 3.2 (Triangle Inequality)[2] *For any vectors p, q in an inner product space $\mathcal{P}$, we have $\|p + q\| \leq \|p\| + \|q\|$.*

Now that we have an inner product space, we can speak of orthogonal polynomials. We call two polynomials p, q *orthogonal with respect to the weight*[3] w provided $\langle p, q \rangle_w = 0$.

---

[2]This is also called the Minkowski inequality.

[3]Although the weight function is important, often we tend not to stress it and call the two polynomials simply *orthogonal*. We also use the simplified notation $\langle \cdot, \cdot \rangle$.

Suppose now that we have a basis $\mathcal{B} = \{\phi_n\}_{n=0}^{\infty}$ for $\mathcal{P}$ consisting of orthogonal polynomials where ϕ_n is of degree n. We can easily see that such a basis exists by beginning with a set of monomials $\{x^n\}_{n=0}^{\infty}$ and, using the Gram-Schmidt process, to come up with the $\{\phi_n\}_{n=0}^{\infty}$: We leave the first member $x^0 = 1$ of the basis as is. That is, $\phi_0 = 1$. Then, from the second member $x^1 = x$, we subtract its component along the first member to get $\phi_1 = x - \frac{\langle x, 1 \rangle}{\langle 1, 1 \rangle} \cdot 1$, so that the second member of our new basis is orthogonal to the first. We continue in this fashion, at each step subtracting from x^n its components along the first $n - 1$ orthogonalized basis members. This creates an orthogonal basis for $\mathcal{P}$.

Now since $\mathcal{B}$ is a basis, given any polynomial $q \in \mathcal{P}$ of degree k we can write[4], for some scalars c_n,

$$q = \sum_{n=0}^{\infty} c_n \phi_n \,, \tag{3.2}$$

i.e., we can expand q in terms of the ϕ_n. Since each polynomial ϕ_n in $\mathcal{B}$ has degree n, we can clearly write any q in terms of the set $\mathcal{B}_k = \{\phi_n\}_{n=0}^{k}$. Indeed, the space of all polynomials of degree k or less has dimension $k + 1$, and the linearly independent set $\mathcal{B}_k$ has $k + 1$ elements. Using the uniqueness of the expansion (3.2), we can see that

$$c_n = 0 \text{ for all } n > k = \deg q\,. \tag{3.3}$$

Since the coefficients are given by

$$c_n = \frac{\langle q, \phi_n \rangle}{\langle \phi_n, \phi_n \rangle}\,,$$

we have the following result.

Proposition 3.3 *Given any polynomial q and any set of orthogonal polynomials $\{\phi_n\}_{n=0}^{\infty}$, where ϕ_n has degree n,*

$$\langle \phi_n, q \rangle_w = \int_I \phi_n q \, w \, dx = 0, \text{ for all } n > \deg q.$$

---

[4]We do not need to worry about the convergence of the following series — we will see in a moment that it is actually a finite sum.

Since we will come across expressions of the form $\langle p, p \rangle$ frequently, let us define the *norm of p with respect to the weight $w(x)$* as

$$\|p\|_w \overset{\text{def}}{=} \sqrt{\langle p, p \rangle_w} .$$

Keeping in mind the way we constructed our set of orthogonal polynomials $\{\phi_n\}_{n=0}^{\infty}$ (requiring ϕ_0 to have degree 0 — i.e., requiring ϕ_0 to be the constant polynomial $\phi_0 = a_0$ — and choosing each successive ϕ_n to be of degree n and orthogonal to the previous n members) the following result should not surprise the reader.

Proposition 3.4 *Any two orthogonal bases of polynomials $\{\phi_n\}_{n=0}^{\infty}$ and $\{\psi_n\}_{n=0}^{\infty}$ for which $\deg \phi_n = \deg \psi_n = n$, must be the same up to some nonzero multiplicative factors.*

Proof Since the $\{\phi_n\}_{n=0}^{\infty}$ constitute a basis, we can write any ψ_m as, using (3.3),

$$\psi_m = \sum_{n=0}^{m} c_n \phi_n$$

for some constants c_n given by

$$c_n = \frac{1}{\|\phi_n\|^2} \langle \psi_m, \phi_n \rangle .$$

By Proposition 3.3, $c_n = 0$ for all $n \neq m$. We are thus left with

$$\phi_m = c_m \psi_m ,$$

and we are done.   ∎

3.2   The Recurrence Formula

We can find a useful recursive relationship between any three consecutive orthogonal polynomials $\phi_{n+1}, \phi_n, \phi_{n-1}$. In what follows, we will let k_n, ℓ_n denote the coefficients of the x^n term and the x^{n-1} term in ϕ_n, respectively.

Proposition 3.5 *Given any set $\mathcal{B} = \{\phi_n\}_{n=0}^{\infty}$ of orthogonal polynomials ϕ_n, such that ϕ_n has degree n, then*

$$\phi_{n+1} - (A_n x + B_n)\phi_n + C_n \phi_{n-1} = 0, \tag{3.4}$$

where

$$A_n = \frac{k_{n+1}}{k_n}, \tag{3.5a}$$

$$B_n = A_n \left(\frac{\ell_{n+1}}{k_{n+1}} - \frac{\ell_n}{k_n} \right), \tag{3.5b}$$

$$C_n = \frac{A_n}{A_{n-1}} \frac{\|\phi_n\|^2}{\|\phi_{n-1}\|^2}, \tag{3.5c}$$

and $C_0 = 0$.

Proof From the formula given for A_n in (3.5), we see that $\phi_{n+1} - A_n x \phi_n$ must be a polynomial of degree n since the leading term of ϕ_{n+1} is canceled. Since the set $\mathcal{B}$ is a basis, we can thus write

$$\phi_{n+1} - A_n x \phi_n = \sum_{j=0}^{n} \gamma_j \phi_j,$$

using (3.3). Taking the inner product of each side with $\phi_{j'}$, where $0 \leq j' \leq n$, we find

$$\gamma_{j'} = \frac{1}{\|\phi_{j'}\|^2} \left(\langle \phi_{n+1}, \phi_{j'} \rangle - A_n \langle x\phi_n, \phi_{j'} \rangle \right),$$

using the orthogonality of $\mathcal{B}$ and the linearity of the inner product. From the definition of the inner product (3.1), we see that the above equation is the same as

$$\gamma_{j'} = \frac{1}{\|\phi_{j'}\|^2} \left(\langle \phi_{n+1}, \phi_{j'} \rangle - A_n \langle \phi_n, x\phi_{j'} \rangle \right).$$

Since $j' \neq n + 1$, the first inner product in the above equation vanishes. Since $\deg(x\phi_{j'}) = \deg(\phi_{j'}) + 1 = j' + 1$, we can use

Proposition 3.3 to determine that the second inner product above is zero unless $j' = n - 1$ or $j' = n$. If we rename $\gamma_n = B_n$ and $\gamma_{n-1} = -C_n$, we have

$$\phi_{n+1} - (A_n\, x + B_n)\phi_n + C_n\phi_{n-1} = 0\,,$$

as in (3.4), with

$$B_n = \frac{-A_n}{\|\phi_n\|^2}\, \langle \phi_n, x\phi_n \rangle, \quad C_n = \frac{A_n}{\|\phi_{n-1}\|^2}\, \langle \phi_n, x\phi_{n-1} \rangle, \qquad (3.6)$$

and it remains to compute the coefficients B_n, C_n. We will start with C_n, rewriting $x\phi_{n-1} = x(k_{n-1}x^{n-1} + \cdots)$ as

$$k_{n-1}x^n + \text{lower order terms} = \frac{k_{n-1}}{k_n}\,(k_n x^n + \text{lower order terms})$$

$$= \frac{k_{n-1}}{k_n}\,(\phi_n + \text{lower order terms})\,.$$

Since the lower order terms "die" in the inner product with ϕ_n, (3.6) implies

$$C_n = \frac{A_n}{\|\phi_{n-1}\|^2}\, \frac{k_{n-1}}{k_n}\, \|\phi_n\|^2 = \frac{A_n}{A_{n-1}}\, \frac{\|\phi_n\|^2}{\|\phi_{n-1}\|^2}\,,$$

as required. We calculate B_n in a similar way, rewriting $x\phi_n = k_n x^{n+1} + \ell_n x^n + \cdots$ as

$$x\phi_n = \frac{k_n}{k_{n+1}}\left[k_{n+1}x^{n+1} + \ell_{n+1}x^n - \left(\ell_{n+1} - \frac{k_{n+1}\ell_n}{k_n}\right) x^n + \text{L.O.T.} \right]$$

$$= \frac{k_n}{k_{n+1}}\phi_{n+1} + \left(\frac{\ell_n}{k_n} - \frac{\ell_{n+1}}{k_{n+1}}\right) k_n x^n + \text{L.O.T.}$$

$$= \frac{k_n}{k_{n+1}}\phi_{n+1} + \left(\frac{\ell_n}{k_n} - \frac{\ell_{n+1}}{k_{n+1}}\right) \phi_n + \text{L.O.T.}\,,$$

where L.O.T. is an acronym which stands for "lower order terms". Only the ϕ_n term will survive the inner product with ϕ_n, and (3.6) implies

$$B_n = \frac{-A_n}{\|\phi_n\|^2}\left(\frac{\ell_n}{k_n} - \frac{\ell_{n+1}}{k_{n+1}}\right) \|\phi_n\|^2 = A_n\left(\frac{\ell_{n+1}}{k_{n+1}} - \frac{\ell_n}{k_n}\right),$$

and we are done.   ∎

We see that if we know the coefficients A_n, B_n, C_n then once we have chosen two consecutive members of our set of orthogonal polynomials, the rest are determined.

3.3   The Rodrigues Formula

Until now, we spoke of the arbitrary interval I. For the remainder of our discussion, we will be most concerned with the interval $[-1, 1]$.

Consider the functions $\psi_n(x)$ defined on $[-1, 1]$ for $n = 0, 1, \ldots$ as

$$\psi_n(x) = \frac{1}{w(x)} \left(\frac{d}{dx} \right)^n \left[w(x)(1 - x^2)^n \right] . \tag{3.7}$$

For arbitrary choice of the weight function w, we cannot say whether these functions are polynomials nor whether they are orthogonal with respect to w. In an effort to force the ψ_n to be polynomials, we can start by looking at ψ_1 and requiring it to be a polynomial of degree 1,

$$\psi_1(x) = \frac{w'(x)}{w(x)}(1 - x^2) - 2x \overset{\text{req}}{=} ax + b .$$

This creates a differential equation, which we can write as

$$\frac{w'}{w} = \frac{(a + 2)x + b}{(1 + x)(1 - x)} .$$

Integrating,

$$\ln w = \int \frac{(a + 2)x + b}{(1 + x)(1 - x)} \, dx + \text{const.}$$

Decomposing the integrand into partial fractions to compute the integral, we find

$$\ln w = -\frac{a + b + 2}{2} \ln (1 - x) + \frac{b - a - 2}{2} \ln (1 + x) + \text{const.} ,$$

or,

$$w(x) = \text{const.} \times (1 - x)^\alpha (1 + x)^\beta ,$$

where α, β are the coefficients from the previous expression. Clearly, the arbitrary constant in the above expression cancels when $w(x)$ is inserted into (3.7), so we will take it to be unity. We will also require $\alpha, \beta > -1$ so that we can integrate any polynomial with respect to this weight. For our purposes, we will see that these restrictions on $w(x)$ are all we need.

Proposition 3.6 *Let $w(x) = (1-x)^\alpha(1+x)^\beta$ with $\alpha, \beta > -1$. Then (3.7) defines a set of polynomials $\{\psi_n\}_{n=0}^\infty$ where $\deg \psi_n = n$.*

In what follows, we will use $(k)_\ell$ to denote the *falling factorial*,

$$(k)_\ell \stackrel{\text{def}}{=} k(k-1)\cdots(k-\ell+1) \text{ for each } \ell \in \mathbb{N},$$

$$(k)_0 \stackrel{\text{def}}{=} 1,$$

and $\binom{n}{k}$ to denote "n choose k"

$$\binom{n}{k} \stackrel{\text{def}}{=} \frac{n!}{k!(n-k)!}. \tag{3.8}$$

Proof We can see this using the *Leibnitz rule* for differentiating products,

$$\left(\frac{d}{dx}\right)^n (fg) = \sum_{k=0}^n \binom{n}{k} \frac{d^k f}{dx^k} \frac{d^{n-k}g}{dx^{n-k}},$$

which can be proved easily by induction.

With the weight $w(x)$ given previously, (3.7) becomes

$$\psi_n(x) = (1-x)^{-\alpha}(1+x)^{-\beta}\left(\frac{d}{dx}\right)^n\left[(1-x)^{\alpha+n}(1+x)^{\beta+n}\right]$$

$$= \sum_{k=0}^n \binom{n}{k}\left[(1-x)^{-\alpha}\left(\frac{d}{dx}\right)^k(1-x)^{n+\alpha}\right] \times$$

$$\left[(1+x)^{-\beta}\left(\frac{d}{dx}\right)^{n-k}(1+x)^{n+\beta}\right]$$

$$= \sum_{k=0}^n \binom{n}{k}\left[(-1)^k(n+\alpha)_k(1-x)^{n-k}\right]\left[(n+\beta)_{n-k}(1+x)^k\right].$$

We can see that ψ_n is a polynomial. Let us examine the leading term in ψ_n. We see from the last line in the above equation that

$$\psi_n(x) = \sum_{k=0}^{n} \binom{n}{k}(-1)^k(n+\alpha)_k(-x)^{n-k}(n+\beta)_{n-k}(x)^k \; + \; \text{L.O.T.}$$

$$= (-1)^n x^n \sum_{k=0}^{n} \binom{n}{k}(n+\alpha)_k(n+\beta)_{n-k} \; + \; \text{L.O.T.}$$

Since the sum in the last line of the above equation is strictly positive, we see that the x^n term in ψ_n has a nonvanishing coefficient; i.e., the degree of ψ_n is precisely n. ∎

Proposition 3.7 Let $w(x) = (1-x)^\alpha(1+x)^\beta$, for $\alpha, \beta > -1$. Then the ψ_n defined in (3.7) satisfy

$$\int_{-1}^{1} \psi_n(x)x^k w(x)\,dx = 0, \quad \text{for } 0 \le k < n.$$

Proof Let J be the above integral. Inserting (3.7) for ψ_n and integrating by parts n times,

$$J = \int_{-1}^{1} x^k \left(\frac{d}{dx}\right)^n \left[w(x)(1-x^2)^n\right] dx$$

$$= x^k \left(\frac{d}{dx}\right)^{n-1} \left[w(x)(1-x^2)^n\right] \Big|_{-1}^{1}$$

$$- \int_{-1}^{1} kx^{k-1} \left(\frac{d}{dx}\right)^{n-1} \left[w(x)(1-x^2)^n\right] dx$$

$$= \cdots$$

$$= \sum_{\ell=0}^{k} (-1)^\ell (k)_\ell \, x^{k-\ell} \left(\frac{d}{dx}\right)^{n-\ell-1} \left[w(x)(1-x^2)^n\right] \Big|_{-1}^{1}.$$

Upon inserting the assumed form of $w(x)$, this becomes

$$J = \sum_{\ell=0}^{k} (-1)^{\ell}(k)_{\ell}\, x^{k-\ell} \left(\frac{d}{dx}\right)^{n-\ell-1} \left[(1-x)^{n+\alpha}(1+x)^{n+\beta}\right]\Big|_{-1}^{1}.$$

In each term of the above sum, the factors $(1-x)^{n+\alpha}$ and $(1+x)^{n+\beta}$ keep positive exponents after being operated upon by $\left(\frac{d}{dx}\right)^{n-\ell-1}$ so that J vanishes after we evaluate the sum at -1 and 1.   ∎

Since each ψ_n has degree n and is orthogonal to all polynomials with degree less than n, we have the following result.

Corollary 3.8 *The formula* (3.7) *defines an orthogonal set of polynomials* $\{\psi_n\}_{n=0}^{\infty}$ *known as the* Jacobi polynomials *with* $\deg \psi_n = n$ *for each* n.

We have thus found that (3.7), called the *Rodrigues formula*, with an acceptable weight function $w(x)$, defines a set of orthogonal polynomials. We are free to multiply each polynomial ψ_n by an arbitrary constant without upsetting this fact. The different classical orthogonal polynomials can be defined using the Rodrigues formula by adjusting this constant and the exponents α, β in $w(x)$.

Now, suppose we discover that a set of polynomials $\{\phi_n\}_{n=0}^{\infty}$ such that $\deg \phi_n = n$ is orthogonal with respect to the weight $w(x)$. Can we conclude that these polynomials satisfy the Rodrigues formula (3.7)? Comparing Corollary 3.8 and Proposition 3.4, we can answer "yes." That is, we can define the ϕ_n by the Rodrigues formula with suitable multiplicative constants,

$$\phi_n(x) = \frac{c_n}{w(x)} \left(\frac{d}{dx}\right)^n \left[w(x)(1-x^2)^n\right].$$

3.4 Approximations by Polynomials

The main result in this section will be the Weierstrass approximation theorem, which will allow us to use polynomials to approximate continuous functions on closed intervals. First, we will introduce some ideas that we will use to prove this important theorem.

Given a function $f : [0, 1] \to \mathbb{R}$, we will use $B_n(x; f)$ to denote a *Bernstein polynomial*, defined on the closed interval $[0, 1]$ as

$$B_n(x; f) \stackrel{\text{def}}{=} \sum_{k=0}^{n} \binom{n}{k} f\left(\frac{k}{n}\right) x^k (1-x)^{n-k}$$

for $n = 1, 2, \ldots$. Since $f\left(\frac{k}{n}\right)$ is just a constant, these are clearly polynomials.

We will compute three special cases of the Bernstein polynomials.

Lemma 3.9 *For every natural number n,*

$$B_n(x; 1) = 1, \qquad (3.9)$$

$$B_n(x; x) = x, \qquad (3.10)$$

$$B_n(x; x^2) = x^2 + \frac{x(1-x)}{n}. \qquad (3.11)$$

Proof By the binomial theorem,

$$(x + y)^n = \sum_{k=0}^{n} \binom{n}{k} x^k y^{n-k}. \qquad (3.12)$$

Substituting $y = 1 - x$ gives

$$1 = \sum_{k=0}^{n} \binom{n}{k} x^k (1-x)^{n-k} = B_n(x; 1), \qquad (3.13)$$

proving (3.9). Differentiating (3.12) with respect to x,

$$n(x + y)^{n-1} = \sum_{k=0}^{n} \binom{n}{k} k x^{k-1} y^{n-k},$$

and multiplying by x/n,

$$x(x + y)^{n-1} = \sum_{k=0}^{n} \binom{n}{k} \left(\frac{k}{n}\right) x^k y^{n-k}. \qquad (3.14)$$

Substituting $y = 1 - x$ gives

$$x = \sum_{k=0}^{n} \binom{n}{k} \left(\frac{k}{n}\right) x^k (1-x)^{n-k} = B_n(x; x),$$

proving (3.10). Differentiating (3.14) with respect to x,

$$(x+y)^{n-1} + (n-1)x(x+y)^{n-2} = \sum_{k=0}^{n} \binom{n}{k} \left(\frac{k^2}{n}\right) x^{k-1} y^{n-k},$$

and multiplying by x/n,

$$\left(x^2 + \frac{xy}{n}\right)(x+y)^{n-2} = \sum_{k=0}^{n} \binom{n}{k} \left(\frac{k}{n}\right)^2 x^k y^{n-k}. \tag{3.15}$$

Substituting $y = 1 - x$ gives

$$x^2 + \frac{x(1-x)}{n} = \sum_{k=0}^{n} \binom{n}{k} \left(\frac{k}{n}\right)^2 x^k (1-x)^{n-k} = B_n(x; x^2),$$

proving (3.11).   ∎

We are now well equipped to prove the following result.

Theorem 3.10 (Weierstrass Approximation Theorem) *Let the function $f : [a, b] \to \mathbb{R}$ be continuous, and let $\epsilon > 0$. Then there exists a polynomial $p(x)$ such that $|f(x) - p(x)| < \epsilon$ for all x in the interval $[a, b]$.*

The Weierstrass approximation theorem is really an existence theorem: It says nothing about how to construct the polynomials that approximate the function $f(x)$. However, in the proof given below, we will show that the Bernstein polynomials, for sufficiently large n, work to approximate the function f to the required accuracy. In other words, the proof we have selected to present is a constructive proof.

Perhaps, we need a clarification here. In the Weierstrass approximation theorem the function f is defined on $[a, b]$ while the Bernstein polynomials are defined on $[0, 1]$. Given, a function f with domain $[a, b]$, we can always redefine it using the linear transformation

$$T(x) = (b - a)x + a,$$

to set

$$\tilde{f} = f \circ T : [0, 1] \to \mathbb{R}.$$

If we can find a polynomial p to adequately approximate $\tilde{f}$, we can use the polynomial $p \circ T^{-1} : [a, b] \to \mathbb{R}$ as our required approximation of f. So, from here on, we will write f instead of $\tilde{f}$ for simplicity and assume that our domain is $[0,1]$.

And a final comment before the proof: The reader should notice that the Bernstein polynomials converge uniformly to $f(x)$, i.e., given an $\epsilon > 0$, there is a $n_0 \in \mathbb{N}$ such that

$$|f(x) - B_n(x; f)| < \epsilon,$$

for all $x \in [0, 1]$ and all $n > n_0$.

We are now ready to present the proof.

Proof Let $\epsilon > 0$. Since f is continuous on the closed and bounded interval $[0, 1]$, we know that f is bounded and uniformly continuous. By the definition of boundedness, there exists an M such that

$$|f(x)| < M, \text{ for every } x \in [0, 1],$$

which implies

$$\left|f(x) - f\left(\frac{k}{n}\right)\right| \le |f(x)| + \left|f\left(\frac{k}{n}\right)\right| < 2M, \qquad (3.16)$$

for all $x \in [0, 1]$ and $0 \le k \le n$. By the definition of uniform continuity, there exists a $\delta > 0$ such that

$$\left|f(x) - f\left(\frac{k}{n}\right)\right| < \frac{\epsilon}{2}, \text{ whenever } |x - \frac{k}{n}| < \delta. \qquad (3.17)$$

Now, let

$$E = |f(x) - B_n(x; f)|,$$

and we will estimate this error from above. Using (3.9) and the triangle inequality,

$$E = \left| f(x) - \sum_{k=0}^{n} \binom{n}{k} f\left(\frac{k}{n}\right) x^k (1-x)^{n-k} \right|$$

$$= \left| \sum_{k=0}^{n} \binom{n}{k} \left[f(x) - f\left(\frac{k}{n}\right) \right] x^k (1-x)^{n-k} \right|$$

$$\leq \sum_{k=0}^{n} \binom{n}{k} \left| f(x) - f\left(\frac{k}{n}\right) \right| x^k (1-x)^{n-k}. \qquad (3.18)$$

Splitting up the sum,

$$E \leq \sum_{|x-\frac{k}{n}|<\delta} \binom{n}{k} \left| f(x) - f\left(\frac{k}{n}\right) \right| x^k (1-x)^{n-k}$$

$$+ \sum_{|x-\frac{k}{n}|\geq\delta} \binom{n}{k} \left| f(x) - f\left(\frac{k}{n}\right) \right| x^k (1-x)^{n-k}.$$

Using (3.16) and (3.17),

$$E < \frac{\epsilon}{2} \sum_{|x-\frac{k}{n}|<\delta} \binom{n}{k} x^k (1-x)^{n-k} + 2M \sum_{|x-\frac{k}{n}|\geq\delta} \binom{n}{k} x^k (1-x)^{n-k}$$

$$\leq \frac{\epsilon}{2} \sum_{k=0}^{n} \binom{n}{k} x^k (1-x)^{n-k} + \frac{2M}{\delta^2} \sum_{k=0}^{n} \binom{n}{k} \left(x - \frac{k}{n} \right)^2 x^k (1-x)^{n-k},$$

or

$$E < \frac{\epsilon}{2} + \frac{2M}{\delta^2} \left[x^2 B_n(x; 1) - 2x B_n(x; x) + B_n(x; x^2) \right].$$

Using Lemma 3.9,

$$E < \frac{\epsilon}{2} + \frac{2M}{\delta^2} \left(x^2 - 2x^2 + x^2 + \frac{x(1-x)}{n} \right).$$

Since the function $x(1-x)$ has a maximum value of $1/4$, we can write

$$E < \frac{\epsilon}{2} + \frac{2M}{4n\delta^2} \le \epsilon, \text{ for all } n \ge \frac{M}{\epsilon\delta^2}. \tag{3.19}$$

Thus, we see that the Bernstein polynomials do the job for large enough n, completing the proof. ∎

Although the previous proof is not complicated, it strongly depends on the form of the polynomilas $B_n(x; f)$. So, the reader may ask the reasonable question "How did Bernstein really think of his polynomials in the first place?" The answer is that "He was fond of *Probability Theory* and he was looking for applications." The Bernstein polynomials are thus a premiere example of how imagination can link in unexpected ways very different areas of mathematics. We will attempt to make this connection [4]. Since the readers may not have background in probability theory, we will digress slightly so we cover the necessary concepts (but we will assume that the reader is familiar with the most basics concepts of random variables).

Let X be a random variable that takes only the two values 1 and 0 with probability $\text{Prob}(X = 1) = p$ and $\text{Prob}(X = 0) = 1 - p$ respectively. For example, X may represent a coin which, when it is flipped, gives heads or tails with probability p and $1 - p$ respectively. The expected value (mean value) of the outcome for a coin-flipping is $\mu = p$ and its variance $\sigma^2 = p(1 - p)$.

The flipping of n identical coins (or just the same coin n times) is described by a random variable X which takes the n values $\{0, 1, 2, ..., n\}$. The probability to get k heads in one trial (the simultaneous flipping of n coins) is

$$\text{Prob}(X = k; n) = \binom{k}{n} p^k (1 - p)^{n-k}.$$

This result is known as the *Bernoulli distribution*. The expected value of the outcome for a trial is $\mu = np$ and its variance $\sigma^2 = np(1 - p)$. Also notice that

$$\sum_{k=0}^{n} \text{Prob}(k; n) = 1.$$

One of the most important results in the probability theory is Chebyshev's inequality that places an upper bound on the probability that a chosen value of a random variable will differ by at least a given amount from the expected value. More precisely:

Theorem 3.11 (Chebyshev Inequality) *Let X be a random variable with expected value μ and variance σ. Given a value x of X and a number $\epsilon > 0$,*

$$\mathrm{Prob}\left(|x - \mu| \geq \epsilon\right) \; \leq \; \frac{\sigma^2}{\epsilon^2}.$$

Now consider a sequence of random variables $X_1, X_2, \ldots, X_n, \ldots$ — if you prefer to keep a concrete example in mind, consider them Bernoulli distributions with increasing n. Just to keep things simple, we assume that the distributions have identical expected values μ and variances σ^2. We then define a new sequence of random variables $M_1, M_2, \ldots, M_n, \ldots$ by

$$M_n \; = \; \frac{X_1 + X_2 + \cdots + X_n}{n}.$$

Each member of the sequence has expected value μ but the variance depends on n:

$$\sigma(M_n) \; = \; \frac{\sigma}{\sqrt{n}}.$$

Applying the Chebyshev inequality for the random variable M_n

$$\mathrm{Prob}\left(\left|\frac{x_1 + x_2 + \cdots + x_n}{n} - \mu\right| \geq \epsilon\right) \; \leq \; \frac{\sigma^2}{n\epsilon^2}.$$

In the limit $n \to \infty$,

$$\lim_{n \to \infty} \mathrm{Prob}\left(\left|\frac{x_1 + x_2 + \cdots + x_n}{n} - \mu\right| \geq \epsilon\right) \; = \; 0.$$

This equation is known as the *weak law of large numbers*. Suppose we apply it to coin-flipping for a coin with $\mu = p$. The quantity

$(x_1 + x_2 + \cdots + x_n)/n$ measures the experimental frequency of heads — that is, it is a number $p_k = k/n$ for some k. Then

$$\lim_{n \to \infty} \text{Prob}\left(|p_k - p| \geq \epsilon\right) = 0.$$

This concludes our digression to probability theory.

Now let's see its application to the Bernstein polynomials. Given an n, we partition the interval $[0,1]$ in n subintervals by the introduction of the points

$$x_k = \frac{k}{n}, \quad k = 0, 1, \cdots, n.$$

The Bernstein polynomial of degree n is

$$B_n(x; f) = \sum_{k=0}^{n} f(x_k) \,\text{Prob}(k; n),$$

with probability to get heads equal to x. For large enough n, some of the points may be found near x; the rest are far. Then, the above sum splits in two sums:

$$B_n(x; f) = \sum_{x_m \simeq x} f(x_m) \,\text{Prob}(m; n) + \sum_{|x_k - x| \geq \epsilon} f(x_k) \,\text{Prob}(k; n).$$

As n increases to infinity, the second sum converges to 0 by the law of large numbers and the first sum converges to $f(x)$.

Now that we know polynomials are a good tool for approximating continuous functions, we ask how to find the best such approximation. This is a result from linear algebra. For a function f, we define the best approximation p to be the one that minimizes the norm of the error, i.e., for which $\|f - p\|$ is smallest.

Proposition 3.12 *Let f be a function, and let $\{\phi_k\}_{k=0}^{\infty}$ be an orthogonal set of polynomials with $\deg \phi_n = n$. The polynomial*

$$p_n = \sum_{k=0}^{n} a_k \phi_k, \quad \text{where} \quad a_k = \frac{\langle f, \phi_k \rangle}{\|\phi_k\|^2}, \tag{3.20}$$

is the unique polynomial of degree n that best approximates f, i.e., that minimizes $\|f - p_n\|$.

Proof Let's choose a different polynomial $\tilde{p}_n$ of degree n. We will write $\tilde{p}_n$ as

$$\tilde{p}_n = p_n + q_n, \quad \text{where} \quad q_n = \sum_{k=0}^{n} b_k \phi_k,$$

for some constants b_k not all zero. Then, we can write $\|f - \tilde{p}_n\|^2$ as

$$\|f - p_n - q_n\|^2 = \|f - p_n\|^2 + \|q_n\|^2 - 2\langle f - p_n, q_n \rangle.$$

But

$$\langle f - p_n, q_n \rangle = \sum_{k=0}^{n} b_k \left(\langle f, \phi_k \rangle - \langle p_n, \phi_k \rangle \right) = 0,$$

since each term in parentheses vanishes, as we can see from (3.20). Thus, for any $\tilde{p}_n$,

$$\|f - \tilde{p}_n\|^2 = \|f - p_n\|^2 + \|q_n\|^2 > \|f - p_n\|^2,$$

the error is greater than that of p_n.   ∎

This fact is really just a special case of a more general result from the study of inner product spaces in linear algebra. We will state the more general fact next. The proof is exactly the same.

Proposition 3.13 *Let f be a vector in an inner product space, and let $\{\phi_k\}_{k=0}^{\infty}$ be an orthogonal set of basis vectors. The vector*

$$p_n = \sum_{k=0}^{n} a_k \phi_k, \quad \text{where} \quad a_k = \frac{\langle f, \phi_k \rangle}{\|\phi_k\|^2},$$

is the unique linear combination of the first $n + 1$ basis vectors that best approximates f, i.e., that minimizes $\|f - p_n\|$.

3.5 Hilbert Space and Completeness

Later in the main discussion, we will be interested in expanding an arbitrary function f in an infinite series of the spherical harmonic functions. In this section, we will address the question of when such an expansion is possible. Since we are concerned primarily with spherical harmonics in an arbitrary number of dimensions, we will keep the discussion general, making no reference here to the number of variables on which the functions depend. In what follows we will thus let x denote a vector in $\mathbb{R}^p$, $\mathcal{D}$ denote a subset of $\mathbb{R}^p$, and $d\Omega$ denote the differential volume element of $\mathcal{D}$.

We will restrict our attention to the space of square-integrable functions with domain $\mathcal{D}$ with respect to the weight $w(x)$, i.e., those functions $f : \mathcal{D} \to \mathbb{R}$ for which the integral

$$\int_{\mathcal{D}} f(x)^2 w(x)\, d\Omega$$

exists and is finite. When endowed with the usual operations of function addition and multiplication by real numbers and paired with the inner product

$$\langle f, g \rangle = \int_{\mathcal{D}} f(x)g(x)w(x)\, d\Omega \,,$$

the set of such functions becomes an inner product space. Such a space has an induced norm

$$\|f\| = \sqrt{\langle f, f \rangle}\,.$$

We will now work up to a definition of a Hilbert space.

Definition In a normed space, a sequence $\{x_n\}_{n=0}^{\infty}$ is called *Cauchy* provided that for any $\epsilon > 0$ there exists an N such that

$$\|x_n - x_m\| < \epsilon \text{ for all } n, m \geq N\,.$$

A Cauchy sequence is always convergent. However, it is possible that its limit does not belong to the same space.

Definition A normed space is *complete* provided every Cauchy sequence converges to an element in the space.

Definition A complete inner product space is called a *Hilbert space*.

We will state the following fact without proof. The interested reader may consult [15] for the proof.

Theorem 3.14 *The inner product space of square-integrable functions defined above is a Hilbert space.*

We can restate the question that opened this section as follows. Given a Hilbert space $\mathcal{H}$ and an *orthonormal* set $\{\phi_n\}_{n=0}^{\infty} \subseteq \mathcal{H}$ (which can be obtained from any orthogonal set by dividing each member by its norm so $\|\phi_n\| = 1$), when can we write any arbitrary member $f \in \mathcal{H}$ as a linear combination[5] of the ϕ_n?

Definition An orthonormal set $\{\phi_n\}_{n=0}^{\infty} \subseteq \mathcal{H}$ is called *complete* provided that for each $f \in \mathcal{H}$, there exist scalars $c_1, c_2, \ldots$, such that

$$\lim_{n \to \infty} \left\| f - \sum_{k=0}^{n} c_k \phi_k \right\| = 0. \qquad (3.21)$$

We know from Proposition 3.13 that out of all linear combinations, the combination

$$p_n = \sum_{k=0}^{n} \langle f, \phi_k \rangle \phi_k$$

minimizes $\|f - p_n\|$. Notice that we have now normalized the vectors ϕ_k to unity: $\|\phi_k\| = 1$. Thus, if there exist scalars c_k for which the norm in (3.21) converges to zero, then certainly

$$\lim_{n \to \infty} \left\| f - \sum_{k=0}^{n} \langle f, \phi_k \rangle \phi_k \right\| = 0, \qquad (3.22)$$

---

[5]A linear combination is usually assumed to contain a finite number of terms. We will not need this restriction for our purposes. Here, linear combinations can be finite or infinite sums.

since

$$\|f - \sum_{k=0}^{n}\langle f, \phi_k\rangle\phi_k\| \le \|f - \sum_{k=0}^{n} c_k\phi_k\|, \text{ for every } n.$$

Let us rewrite (3.22) by computing

$$\left\|f - \sum_{k=0}^{n}\langle f_n, \phi_k\rangle\phi_k\right\|^2,$$

which is the same as

$$\|f\|^2 - 2\left\langle f, \sum_{k=0}^{n}\langle f, \phi_k\rangle\phi_k\right\rangle + \left\langle \sum_{k=0}^{n}\langle f, \phi_k\rangle\phi_k, \sum_{\ell=0}^{n}\langle f, \phi_\ell\rangle\phi_\ell\right\rangle.$$

Using the linearity of the inner product, this becomes

$$\|f\|^2 - 2\sum_{k=0}^{n}\langle f, \phi_k\rangle^2 + \sum_{k,\ell=0}^{n}\langle f, \phi_k\rangle\langle f, \phi_\ell\rangle\langle\phi_k, \phi_\ell\rangle.$$

By the orthonormality of the ϕ_n, we have

$$\left\|f - \sum_{k=0}^{n}\langle f_n, \phi_k\rangle\phi_k\right\|^2 = \|f\|^2 - \sum_{k=0}^{n}\langle f, \phi_k\rangle^2. \qquad (3.23)$$

We note that since the norm-squared is non-negative, then

$$\|f\|^2 - \sum_{k=0}^{n}\langle f, \phi_k\rangle^2 \ge 0,$$

or,

$$\sum_{k=0}^{n}\langle f, \phi_k\rangle^2 \le \|f\|^2, \text{ for every } n. \qquad (3.24)$$

This is known as *Bessel's inequality* and tells us that the sum in the left-hand side converges as $n \to \infty$. If the ϕ_k form a complete set, $\|f - \sum_{k=0}^{n}\langle f_n, \phi_k\rangle\phi_k\|^2$ must converge to zero as $n \to \infty$. In this case, from equation (3.23), it follows that Bessel's inequality becomes an equality:

$$\sum_{k=0}^{\infty}\langle f,\phi_k\rangle^2 = \|f\|^2,$$

known as *Parseval's equality*. We have thus arrived at the following conclusion.

Proposition 3.15 *An orthonormal set $\{\phi_n\}_{n=0}^{\infty} \subseteq \mathcal{H}$ is complete if and only if Parseval's equality holds for each $f \in \mathcal{H}$.*

We give one more definition.

Definition A set $\{\phi_n\}_{n=0}^{\infty}$ is *closed* provided

$$\langle f,\phi_n\rangle = 0 \text{ for every } n \qquad \text{implies} \qquad f = 0.$$

Now we are ready to prove the main result of this section.

Theorem 3.16 *The orthonormal set $\{\phi_n\}_{n=0}^{\infty} \subseteq \mathcal{H}$ is complete if and only if it is closed.*

Proof (Sufficient) Let $\{\phi_n\}_{n=0}^{\infty} \subset \mathcal{H}$ constitute a closed orthonormal set and $f \in \mathcal{H}$. We will show that

$$\lim_{n\to\infty}\left(\|f\|^2 - \sum_{k=0}^{n}\langle f,\phi_k\rangle^2\right) = 0,$$

since this guarantees completeness by Proposition 3.15.

Define $g_n = f - \sum_{k=0}^{n}\langle f,\phi_k\rangle\phi_k$, and notice that $\{g_n\}_{n=0}^{\infty}$ is a Cauchy sequence since, if $n > m$,

$$\|g_n - g_m\|^2 = \left\|\sum_{k=m+1}^{n}\langle f,\phi_k\rangle\phi_k\right\|^2$$

$$= \sum_{k,\ell=m+1}^{n}\langle f,\phi_k\rangle\langle f,\phi_\ell\rangle\langle\phi_k,\phi_\ell\rangle$$

$$= \sum_{k=m+1}^{n}\langle f,\phi_k\rangle^2,$$

which can be made arbitrarily small for large enough n, m since the series $\sum_{k=0}^{\infty}\langle f, \phi_k \rangle^2$ converges, by (3.24). Since $\mathcal{H}$ is complete by definition, this Cauchy sequence must converge to some $g \in \mathcal{H}$, i.e.,

$$\lim_{n \to \infty} \|g_n - g\| = 0. \tag{3.25}$$

Now fix j and take $n \in \mathbb{N}_0$ such that $n > j$. By the Cauchy-Schwartz inequality

$$|\langle g, \phi_j \rangle| = |\langle g_n - g, \phi_j \rangle| \leq \|g_n - g\| \|\phi_j\| = \|g_n - g\|.$$

Since this holds for any $n > j$, we can use (3.25) to conclude

$$|\langle g, \phi_j \rangle| \leq \lim_{n \to \infty} \|g_n - g\| = 0,$$

which implies that for arbitrary j, $\langle g, \phi_j \rangle = 0$. Since we assumed the ϕ_j constitute a closed set, we have $g = 0$. Therefore,

$$\lim_{n \to \infty} \left(\|f\|^2 - \sum_{k=0}^{n} \langle f, \phi_k \rangle^2 \right) = \lim_{n \to \infty} \|g_n\|^2 = \lim_{n \to \infty} \|g_n - g\|^2 = 0.$$

(Necessary) Suppose $\{\phi_n\}_{n=0}^{\infty}$ is complete but not closed. Then there exists a function $f \neq 0$ such that $\langle f, \phi_n \rangle = 0$ for every n. Then

$$\lim_{n \to \infty} \left(\|f\|^2 - \sum_{k=0}^{n} \langle f, \phi_k \rangle^2 \right) = \|f\|^2 \neq 0.$$

So, by Proposition 3.15, $\{\phi_n\}_{n=0}^{\infty}$ is not complete — a contradiction. This proves that if an orthonormal set in $\mathcal{H}$ is closed, then it is complete.   ∎

In Chapter 4, we will use this result to show that the set of spherical harmonics form a complete set in the space of square-integrable functions by showing that it is closed.

3.6   Problems

1. In many books the polynomials

$$B_{k,n}(x) = \binom{k}{n} x^k (1-x)^{n-k}, \quad k = 0, 1, \ldots, n, \quad x \in [0,1],$$

are referred to as the *Bernstein polynomials*. Notice that these are the probabilities of the Bernoulli distribution. Hence the property

$$\sum_{k=0}^{n} B_{k,n}(x) = 1.$$

Prove that these polynomials obey the recursion relations

$$B_{k,n}(x) = (1-x) B_{k,n-1}(x) + x B_{k-1,n-1}(x),$$
$$B_{k,n-1}(x) = \frac{n-k}{n} B_{k,n}(x) + \frac{k+1}{n} B_{k+1,n}(x),$$

and that their derivatives satisfy

$$\frac{d}{dx} B_{k,n}(x) = n \left(B_{k-1,n-1}(x) - B_{k,n-1}(x) \right).$$

2. Compute the Bernstein polynomials $B_n(x; x^3)$ for any n.

3. If $f(x)$ has a continuous derivative in [0,1], prove that the sequence of derivatives

$$\frac{d}{dx} B_0(x; f), \quad \frac{d}{dx} B_1(x; f), \quad \ldots, \quad \frac{d}{dx} B_n(x; f), \quad \ldots$$

converges uniformly to $f'(x)$ in [0,1].

Chapter 4

Spherical Harmonics in p Dimensions

We will begin by developing some facts about a special kind of polynomials. We will then define a spherical harmonic to be one of these polynomials with a restricted domain, as hinted at in Section 1.1. After discussing some properties of spherical harmonics, we will introduce the Legendre polynomials. And once we have produced a considerable number of results, we will move on to an application of the material developed to boundary value problems.

4.1 Harmonic Homogeneous Polynomials

Definition A polynomial $H_n(x_1, x_2, \ldots, x_p)$ is *homogeneous of degree n* in the p variables $x_1, x_2, \ldots, x_p$ provided

$$H_n(tx_1, tx_2, \ldots, tx_p) = t^n H_n(x_1, x_2, \ldots, x_p) \ .$$

In the definition of a homogeneous polynomial, let's set $u_i = tx_i$, for all i and differentiate the defining equation with respect to t:

$$\sum_{i=1}^{p} \frac{\partial H_n(u_1, u_2, \ldots, u_p)}{\partial u_i} \frac{du_i}{dt} = n\, t^{n-1} H_n(x_1, x_2, \ldots, x_p),$$

or

$$\sum_{i=1}^{p} \frac{\partial H_n(u_1, u_2, \ldots, u_p)}{\partial u_i} x_i = n\, t^{n-1}\, H_n(x_1, x_2, \ldots, x_p)\,.$$

Finally, we set $t = 1$ to find the following functional equation satisfied by the homogeneous polynomial,

$$\sum_{i=1}^{p} \frac{\partial H_n(x_1, x_2, \ldots, x_p)}{\partial x_i} x_i = n\, H_n(x_1, x_2, \ldots, x_p)\,, \qquad (4.1)$$

known as *Euler's equation*.

The following calculation will be useful in the counting of linearly independent homogeneous polynomials.

Lemma 4.1 *For* $0 < |r| < 1$,

$$\frac{1}{(1-r)^p} = \sum_{j=0}^{\infty} \frac{(p+j-1)!}{j!(p-1)!}\, r^j\,. \qquad (4.2)$$

Proof We will use a counting trick to prove this. Since $0 < |r| < 1$, we can write $(1-r)^{-p}$ in terms of a product of geometric series, i.e.,

$$(1-r)^{-p} = \underbrace{\left(\frac{1}{1-r}\right)\left(\frac{1}{1-r}\right)\cdots\left(\frac{1}{1-r}\right)}_{p \text{ times}}$$

$$= \underbrace{\left(\sum_{n=0}^{\infty} r^n\right)\left(\sum_{n=0}^{\infty} r^n\right)\cdots\left(\sum_{n=0}^{\infty} r^n\right)}_{p \text{ times}}\,.$$

We will compute the *Cauchy product* of the p infinite series. The result will be an infinite series including a constant term and all positive integer powers of r,

$$(1-r)^{-p} = \sum_{j=0}^{\infty} c_j r^j\,.$$

It remains to compute the coefficients c_j. To determine each c_j, we must compute how many r^j's are produced in the Cauchy product, i.e., in how many different ways we produce an r^j in the multiplication.

Note that this computation is equivalent to asking, "In how many different ways can we place j indistinguishable balls into p boxes?" The p boxes correspond to the p series in the product, and choosing to place $k < j$ balls into a certain box corresponds to choosing the r^k term in that series when computing a term of the Cauchy product.

Let us use a diagram to assist in this calculation. We will use a vertical line to denote a division between two boxes and a dot to denote a ball. Each configuration of the j balls in p boxes can thus be represented by a string of j dots and $p-1$ lines (since p boxes require only $p-1$ divisions). For example,

$$\bullet\bullet\,|\,\bullet\,|\,\bullet\,|\,|\,\bullet\bullet\,|\,\bullet\,|\,|\bullet$$

represents one configuration of $j = 8$ balls in $p = 8$ boxes.

Now, the number of ways to arrange j indistinguishable balls in p boxes is the same as the number of distinct arrangements of j dots and $p-1$ lines. This is given by "$p-1+j$ choose j," i.e.,

$$c_j \;=\; \binom{p-1+j}{j} \;=\; \frac{(p-1+j)!}{j!(p-1)!}\,,$$

and the lemma is proved.   ∎

Proposition 4.2 *If $K(p,n)$ denotes the number of linearly independent homogeneous polynomials of degree n in p variables, then*

$$K(p,n) \;=\; \frac{(p+n-1)!}{n!(p-1)!}\,.$$

We will give two proofs of this claim. The first uses a recursive relation obeyed by $K(p,n)$, while the second employs another counting trick.

Proof 1 Let $H_n(x_1, x_2, \ldots, x_p)$ be a homogeneous polynomial of degree n in its p variables. Notice that H_n is a polynomial in x_p of degree at most n. For if H_n contained a power of x_p greater than n, the polynomial could not be homogenous of degree n, since $H_n(\ldots, tx_p)$ would contain a power of t greater than n. Thus we can write,

$$H_n(x_1, x_2, \ldots, x_p) = \sum_{j=0}^{n} x_p^j \, h_{n-j}(x_1, x_2, \ldots, x_{p-1}), \qquad (4.3)$$

where the h_{n-j} are polynomials. Moreover, notice that the h_{n-j} must be homogeneous of degree $n-j$ in their $p-1$ variables. Indeed, using the homogeneity of H_n,

$$t^n \sum_{j=0}^{n} x_p^j h_{n-j}(x_1, \ldots, x_{p-1}) \quad = \quad H_n(tx_1, \ldots, tx_p)$$

$$= \quad \sum_{j=0}^{n} (tx_p)^j h_{n-j}(tx_1, \ldots, tx_{p-1}),$$

which implies that

$$\sum_{j=0}^{n} x_p^j \left[t^{n-j} h_{n-j}(x_1, \ldots, x_{p-1}) - h_{n-j}(tx_1, \ldots, tx_{p-1}) \right] = 0.$$

The expression in brackets must vanish by the linear independence of the x_p^j. Each h_{n-j} can be written in terms of a basis of $K(p-1, n-j)$ homogeneous polynomials of degree $n-j$ in $p-1$ variables, and thus H_n can be written in terms of a basis of

$$K(p, n) = \sum_{j=0}^{n} K(p-1, n-j) = \sum_{j=0}^{n} K(p-1, j)$$

linearly independent elements. We have found a recursive relation that the $K(p, n)$ must satisfy. Now for some $0 < |r| < 1$, let

$$G(p) = \sum_{n=0}^{\infty} r^n K(p, n). \qquad (4.4)$$

Then,

$$G(p) = \sum_{n=0}^{\infty} r^n \sum_{j=0}^{n} K(p-1,j) = \sum_{j=0}^{\infty} K(p-1,j) \sum_{n=j}^{\infty} r^n,$$

or

$$G(p) = \sum_{j=0}^{\infty} K(p-1,j) r^j \sum_{n=0}^{\infty} r^n = \frac{1}{1-r} \sum_{j=0}^{\infty} r^j K(p-1,j),$$

so

$$G(p) = \frac{G(p-1)}{1-r}.$$

Using an inductive argument, we can show that

$$G(p) = \frac{G(1)}{(1-r)^{p-1}}.$$

By noticing that $K(1,n) = 1$, since every homogeneous polynomial of degree n in one variable can be written as cx_1^n, we see that

$$G(1) = \sum_{j=0}^{\infty} r^n = \frac{1}{1-r}.$$

Thus,

$$G(p) = (1-r)^{-p} = \sum_{n=0}^{\infty} \frac{(p+n-1)!}{n!(p-1)!} r^n,$$

by the above lemma. Comparing this to (4.4) then proves the theorem. ∎

Proof 2 Using reasoning similar to that used in the above proof, we see that every *monomial* (i.e., product of variables) in an n-th degree homogeneous polynomial must be of degree n. For example, a fourth degree homogeneous polynomial in the variables x, y can have an $x^2 y^2$ but not an $x^2 y$ term.

To uniquely determine such a polynomial, we must give the coefficient of every possible n-th degree monomial. So to find $K(p, n)$, we must find out how many possible n-th degree monomials there are in p variables. But this is equivalent to asking, "In how many ways can we place n indistinguishable balls into p boxes?" The p boxes represent the p variables from which we can choose, and placing $j < n$ balls into a certain box corresponds to choosing to raise that variable to the j power. For example, using the same notation as in the proof of Lemma 4.1, the string

$$\bullet \, \bullet \, | \, \bullet \, | \, \bullet \, |$$

could represent the $w^2 x^1 y^1 z^0$ term in a fourth-degree homogeneous polynomial. As in the above lemma, the number of such arrangements is

$$K(p,n) \;=\; \binom{n+p-1}{n} = \frac{(n+p-1)!}{n!(p-1)!},$$

proving the theorem.   ∎

Definition A polynomial $q(x_1, x_2, \ldots, x_p)$ is *harmonic* provided it satisfies the Laplace equation

$$\Delta_p q \;=\; 0. \tag{4.5}$$

The following property of combinations will be used to prove the theorem that follows it.

Lemma 4.3 *If $k, \ell \in \mathbb{N}$, then*

$$\binom{k}{\ell} \;=\; \frac{k}{\ell}\binom{k-1}{\ell-1}.$$

Proof Just compute:

$$\frac{k}{\ell}\binom{k-1}{\ell-1} \;=\; \frac{k}{\ell} \cdot \frac{(k-1)!}{(\ell-1)!(k-\ell)!} \;=\; \frac{k!}{\ell!(k-\ell)!} \;=\; \binom{k}{\ell},$$

as required.   ∎

Theorem 4.4 *If $N(p, n)$ denotes the number of linearly independent homogeneous harmonic polynomials of degree n in p variables, then*

$$N(p, n) = \frac{2n + p - 2}{n} \binom{n + p - 3}{n - 1}.$$

Proof Let H_n be a homogeneous harmonic polynomial of degree n in p variables. As in (4.3), we write

$$H_n(x_1, x_2, \ldots, x_p) = \sum_{j=0}^{n} x_p^j \, h_{n-j}(x_1, x_2, \ldots, x_{p-1}),$$

and we operate on H_n with the Laplace operator. Since H_n is harmonic,

$$
\begin{aligned}
0 = \Delta_p H_n &= \left(\frac{\partial^2}{\partial x_p^2} + \Delta_{p-1} \right) H_n \\
&= \sum_{j=2}^{n} j(j-1) x_p^{j-2} h_{n-j} + \sum_{j=0}^{n} x_p^j \Delta_{p-1} h_{n-j},
\end{aligned}
$$

so

$$0 = \sum_{j=0}^{n} x_p^j \left[(j+2)(j+1) h_{n-j-2} + \Delta_{p-1} h_{n-j} \right].$$

where we define $h_{-1} = h_{-2} = 0$. Since the x_p^j are linearly independent, each of the coefficients must vanish,

$$
\begin{aligned}
2h_{n-2} + \Delta_{p-1} h_n &= 0, \\
6h_{n-3} + \Delta_{p-1} h_{n-1} &= 0, \\
&\;\;\vdots \\
n(n-1)h_0 + \Delta_{p-1} h_2 &= 0, \\
\Delta_{p-1} h_1 &= 0, \\
\Delta_{p-1} h_0 &= 0.
\end{aligned}
\tag{4.6}
$$

We have found a recursive relationship that the h_{n-j} must obey. Thus, choosing h_n and h_{n-1} determines the rest of the h_{n-j}. By Theorem 4.2, h_n can be written in terms of $K(p-1,n)$ basis polynomials and h_{n-1} can be written in terms of $K(p-1,n-1)$ basis polynomials. Thus we must give $K(p-1,n)+K(p-1,n-1)$ coefficients to determine h_n, h_{n-1} which thereby determine H_n. Therefore,

$$\begin{aligned} N(p,n) &= K(p-1,n)+K(p-1,n-1) \\ &= \binom{p+n-2}{n}+\binom{p+n-3}{n-1}. \end{aligned}$$

By the above lemma, we can write this as

$$\begin{aligned} N(p,n) &= \frac{p+n-2}{n}\binom{p+n-3}{n-1}+\binom{p+n-3}{n-1} \\ &= \frac{2n+p-2}{n}\binom{n+p-3}{n-1}, \end{aligned}$$

thus the theorem is proved. ∎

Notice that the previous result can also be written as

$$N(p,n) = \binom{n+p-1}{n}-\binom{n+p-3}{n-2} = K(p,n)-K(p,n-2).$$

The interpretation of this formula is very easy: Δ is a second-order (linear) differential operator mapping polynomials H_n to H_{n-2}, as we have seen explicitly. The above equation states that the number of linearly independent, harmonic, homogeneous polynomials of degree n is just the difference in the dimensions of the two spaces corresponding to H_n and H_{n-2}. But we could have anticipated this results, since we know from linear algebra that for a linear operator, the dimension of its domain (in this case $K(p,n)$) equals the dimensions of its range (here, $K(p,n-2)$) plus the dimension of its null space (which is $N(p,n)$ in this case).

In what follows, x will always denote the vector $(x_1, x_2, \ldots, x_p)$, r or $|x|$ will denote its norm $\sqrt{x_1^2+x_2^2+\cdots+x_p^2}$, and ξ will denote the vector $(\xi_1, \xi_2, \ldots, \xi_p)$ having unit norm. Keep in mind that x, ξ represent vectors while the x_j, ξ_j denote their components. Also, H_n will always denote a harmonic homogeneous polynomial. Lastly, in the remainder of this chapter, we will often suppress explicit reference to the number of dimensions in which we work. It should be assumed that we deal with p dimensions unless stated otherwise.

4.2 Spherical Harmonics and Orthogonality

We are now ready to introduce the spherical harmonics. First, notice that

$$H_n(x) = H_n(r\xi) = r^n H_n(\xi). \tag{4.7}$$

In p-dimensional spherical coordinates, the radial dependence and the angular dependence of the functions H_n can be separated.

Definition A *spherical harmonic of degree* n, denoted $Y_n(\xi)$, is a harmonic homogeneous polynomial of degree n in p variables restricted to the unit $(p-1)$-sphere. In other words, Y_n is the map

$$Y_n : S^{p-1} \to \mathbb{E}, \text{ given by } Y_n(\xi) = H_n(\xi) \text{ for every } \xi \in S^{p-1}$$

for some harmonic homogeneous polynomial H_n. We can write $Y_n = H_n|_{S^{p-1}}$.

In Chapter 1, we introduced functions Y (which we claimed were spherical harmonics) that turned out to be eigenfunctions of the angular part of the Laplace operator. Using Proposition 2.5, we can show that spherical harmonics are indeed eigenfunctions of $\Delta_{S^{p-1}}$.

Proposition 4.5

$$\Delta_{S^{p-1}} Y_n = n(2 - p - n) Y_n. \tag{4.8}$$

Proof Let H_n be a harmonic homogeneous polynomial and Y_n its associated spherical harmonic. As in (4.7), we have $H_n(x) = r^n Y_n(\xi)$. Then, using Proposition 2.5,

$$0 = \Delta_p(r^n Y_n) = n(n-1)r^{n-2}Y_n + \frac{p-1}{r} nr^{n-1}Y_n + \frac{1}{r^2} r^n \Delta_{S^{p-1}} Y_n.$$

Rearranging,

$$r^{n-2}\left[\Delta_{S^{p-1}}Y_n + n(n+p-2)Y_n\right] = 0,$$

which implies

$$\Delta_{S^{p-1}}Y_n = n(2-p-n)Y_n,$$

as sought.   ∎

Remark In three dimensions (4.8) becomes

$$\left[\frac{1}{\sin\theta}\frac{\partial}{\partial\theta}\left(\sin\theta\frac{\partial}{\partial\theta}\right) + \frac{1}{\sin^2\theta}\frac{\partial^2}{\partial\phi^2}\right]Y_\ell = -\ell(\ell+1)Y_\ell.$$

As promised in Subsection 1.2, this allows us to show that three-dimensional spherical harmonics carry a definite amount of quantum mechanical angular momentum. Referring to (1.13),

$$\hat{\vec{L}}^2 Y_\ell = \hbar^2\ell(\ell+1)Y_\ell.$$

Remark Since the spherical harmonic $Y_n(\xi)$ is defined as the restriction of some $H_n(x)$ to the unit sphere, $Y_n(\xi)$ must also be homogeneous. However, $Y_n(t\xi)$ is not always defined since the domain of the spherical harmonic is S^{p-1}. In fact, it is only defined for $|t\xi| = 1$, and since $|\xi| = 1$, we see that we can only write $Y_n(t\xi)$ for $t = \pm 1$. The case $t = 1$ is trivial, but the case $t = -1$ tells us the parity of $Y_n(\xi)$. Since $Y_n(-\xi) = (-1)^n Y_n(\xi)$, we see that the transformation $\xi \longmapsto -\xi$ sends $Y_n \longmapsto -Y_n$ if n is odd and leaves Y_n invariant if n is even.

We now come to the main result of this section, which we hinted at in Section 1.1.

Theorem 4.6 *Let $Y_n(\xi), Y_m(\xi)$ be two spherical harmonics. Then*

$$\int_{S^{p-1}} Y_n(\xi)Y_m(\xi)\,d\Omega_{p-1} = 0, \quad \text{if } n \neq m.$$

That is, spherical harmonics of different degrees are orthogonal over the sphere.

Proof Let us perform the computation using the harmonic homogeneous polynomials H_n and H_m where $Y_n = H_n|_{S^{p-1}}$ and $Y_m = H_m|_{S^{p-1}}$.

Now, we start with the divergence theorem in p dimensions for a vector A,

$$\int_{B^p} \nabla_p \cdot A(x)\, d^p x = \int_{S^{p-1}} A(\xi) \cdot \xi\, d\Omega_{p-1}\,,$$

apply it twice for the vectors $H_n \nabla H_m$ and $H_m \nabla H_n$, and subtract the results[1]:

$$\int_{B^p} \nabla_p \cdot [H_n(x)\nabla_p H_m(x) - H_m(x)\nabla_p H_n(x)]\, d^p x$$

$$= \int_{S^{p-1}} [H_n(\xi)\nabla_p H_m(\xi) - H_m(\xi)\nabla_p H_n(\xi)] \cdot \xi\, d\Omega_{p-1}\,,$$

or

$$0 = \int_{S^{p-1}} [H_n(\xi)\nabla_p H_m(\xi) - H_m(\xi)\nabla_p H_n(\xi)] \cdot \xi\, d\Omega_{p-1}\,, \qquad (4.9)$$

where we used the property

$$\nabla_p \cdot (H_n \nabla_p H_m) = \nabla_p H_n \cdot \nabla_p H_m + H_n \Delta_p H_m\,,$$

and the fact $\Delta_p H_m = \Delta_p H_n = 0$.

---

[1]Incidentally, we point out that the identity thus obtained for any two functions f, g

$$\int_{B^p} (f\Delta_p g - g\Delta_p f)\, d^p x = \int_{S^{p-1}} (f\nabla_p g - g\nabla_p f) \cdot \xi\, d\Omega_{p-1}\,,$$

is known as *Green's theorem*. We may also observe that $\nabla_p f \cdot \xi$ is the directional derivative of f along ξ and write

$$\int_{B^p} (f\Delta_p g - g\Delta_p f)\, d^p x = \int_{S^{p-1}} \left(f\frac{\partial g}{\partial \xi} - g\frac{\partial f}{\partial \xi} \right) d\Omega_{p-1}\,.$$

Euler's equation (4.1) for a homogeneous polynomial,

$$\sum_{j=1}^{p} \frac{\partial H_n(\xi)}{\partial \xi_j} \xi_j = n H_n(\xi),$$

can be written in the form

$$\nabla_p H_n(\xi) \cdot \xi = n H_n(\xi).$$

With the help of this result, equation (4.9) takes the form

$$(m-n) \int_{S^{p-1}} H_n(\xi) H_m(\xi) d\Omega_{p-1} = 0.$$

But the integral is carried out over S^{p-1} where $Y_n = H_n$ and $Y_m = H_m$. This equation is thus equivalent to

$$(m-n) \int_{S^{p-1}} Y_n(\xi) Y_m(\xi) d\Omega_{p-1} = 0.$$

By hypothesis, $n \neq m$; therefore Y_n, Y_m are orthogonal over the sphere. ∎

Given a set of $N(p,n)$ linearly independent spherical harmonics of degree n, we can use the Gram-Schmidt orthonormalization procedure to produce an orthonormal set of spherical harmonics, i.e., a set

$$\{Y_{n,i}(\xi)\}_{i=1}^{N(p,n)} \quad \text{with} \quad \int_{S^{p-1}} Y_{n,i}(\xi) Y_{n,j}(\xi) d\Omega_{p-1} = \delta_{ij}, \quad (4.10)$$

where

$$\delta_{ij} = \begin{cases} 1 & \text{if } i = j, \\ 0 & \text{if } i \neq j, \end{cases}$$

is the Kronecker delta. For the remainder of this chapter, unless indicated otherwise, we will let $Y_{n,i}(\xi)$ denote an n-th degree spherical harmonic belonging to an orthonormal set of $N(p,n)$ such functions, as in (4.10).

In what follows, let R be an orthogonal matrix that acts on ξ as a rotation of coordinates. Notice that since the integration is taken over the entire sphere in (4.10), the orthonormal set of spherical harmonics remains orthonormal in a rotated coordinate frame. That is,

$$\int_{S^{p-1}} Y_{n,i}(R\xi)\, Y_{n,j}(R\xi)\, d\Omega_{p-1} \;=\; \delta_{ij}\,. \tag{4.11}$$

Proposition 4.7 *If $Y_n(\xi)$ is a spherical harmonic of degree n, then $Y_n'(\xi) = Y_n(R\xi)$ is also a spherical harmonic of degree n, for any rotation matrix R.*

Proof Let $Y_n(\xi)$ be a spherical harmonic of degree n. Then there exists a harmonic homogeneous polynomial $H_n(x)$ of degree n such that $Y_n = H_n|_{S^{p-1}}$. Denote $H_n'(x) = H_n(Rx)$. We claim that $Y_n' = H_n'|_{S^{p-1}}$. To see this, first notice that $H_n'(x)$ is a polynomial in $x_1, x_2, \ldots, x_p$. Indeed, $H_n'(x) = H_n(Rx)$ is a linear combination of powers of the $\sum_{j=1}^{p} R_{ij} x_j$ and thus a linear combination of powers of the x_j. Next, notice that H_n' is homogeneous of degree n,

$$H_n'(tx) \;=\; H_n(tRx) = t^n\, H_n(Rx) \;=\; t^n\, H_n'(x)\,.$$

Finally, notice that H_n' is harmonic, by Proposition 2.1. Restricting H_n' to the unit sphere thus gives a spherical harmonic of degree n, $Y_n'(\xi)$. ∎

Since the set $\{Y_{n,i}(\xi)\}_{i=1}^{N(p,n)}$ in (4.10) is a maximal linearly independent set of spherical harmonics of degree n, it serves as a basis for all such functions. We have just shown that $Y_{n,j}(R\xi)$ is a spherical harmonic of degree n provided $Y_{n,j}(\xi)$ is as well. Thus, we can write $Y_{n,j}(R\xi)$ in terms of the basis functions:

$$Y_{n,j}(R\xi) \;=\; \sum_{\ell=1}^{N(p,n)} C_{\ell j}\, Y_{n,\ell}(\xi)\,.$$

Using this expression to rewrite the integral in (4.11), we can show that the matrix C defined in the above expression is orthogonal. Indeed,

$$
\begin{aligned}
\delta_{ij} &= \int_{\xi \in S^{p-1}} \left(\sum_{k=1}^{N(p,n)} C_{ki} Y_{n,k}(\xi) \right) \left(\sum_{\ell=1}^{N(p,n)} C_{\ell j} Y_{n,\ell}(\xi) \right) d\Omega_{p-1} \\
&= \sum_{k,\ell=1}^{N(p,n)} C_{ki} C_{\ell j} \int_{\xi \in S^{p-1}} Y_{n,k}(\xi) Y_{n,\ell}(\xi) d\Omega_{p-1} \\
&= \sum_{k,\ell=1}^{N(p,n)} C_{ki} C_{\ell j} \delta_{k\ell} = \sum_{k=1}^{N(p,n)} C_{ik}^{t} C_{kj} .
\end{aligned}
$$

For the following discussion, let ξ, η be two unit vectors. Let us consider the function given by

$$
F_n(\xi, \eta) = \sum_{j=1}^{N(p,n)} Y_{n,j}(\xi) Y_{n,j}(\eta). \tag{4.12}
$$

Lemma 4.8 *The function F_n defined above is invariant under a rotation of coordinates.*

Proof Let R be a rotation matrix. Then, using the orthogonal matrix C discussed above,

$$
\begin{aligned}
F_n(R\xi, R\eta) &= \sum_{j=1}^{N(p,n)} Y_{n,j}(R\xi) Y_{n,j}(R\eta) \\
&= \sum_{j=1}^{N(p,n)} \left(\sum_{\ell=1}^{N(p,n)} C_{\ell j} Y_{n,\ell}(\xi) \right) \left(\sum_{m=1}^{N(p,n)} C_{mj} Y_{n,m}(\eta) \right),
\end{aligned}
$$

so

$$F_n(R\xi, R\eta) = \sum_{\ell,m=1}^{N(p,n)} Y_{n,\ell}(\xi)\, Y_{n,m}(\eta) \left(\sum_{j=1}^{N(p,n)} C_{\ell j}\, C_{jm}^t \right)$$

$$= \sum_{\ell=1}^{N(p,n)} Y_{n,\ell}(\xi)\, Y_{n,\ell}(\eta) \;=\; F_n(\xi,\eta)\,,$$

as claimed.   ∎

Since the dot product $\langle \xi, \eta \rangle$ is also invariant under the rotation R, this suggests that $F_n(\xi, \eta)$ could be a function of $\langle \xi, \eta \rangle$ alone. In fact, this is the case, as shown in the following lemma.

Lemma 4.9 *If F_n is defined as in (4.12), then $F_n(\xi, \eta) = p(\langle \xi, \eta \rangle)$ where $p(t)$ is a polynomial.*

Proof A rotation of coordinates leaves $\langle \xi, \eta \rangle$ invariant and, by the above lemma, does not change $F_n(\xi, \eta)$ either. For some $-1 \le t \le 1$, there exists a rotation R that sends

$$\xi \xmapsto{R} \xi' = (t, \sqrt{1 - t^2}, 0, \ldots, 0)\,,$$

$$\eta \xmapsto{R} \eta' = (1, 0, \ldots, 0)\,.$$

To see this, just rotate coordinates so that η points along the x_1-axis. Then rotate coordinates around the x_1-axis until the component of ξ orthogonal to η points along the x_2-axis.

Notice that $\langle \xi, \eta \rangle = \langle \xi', \eta' \rangle = t$. Since $F_n(\xi, \eta)$ is a sum of products of spherical harmonics, which are each polynomials in the components of ξ or η, F_n is a polynomial, say p, in the components of its arguments, i.e.,

$$F_n(\xi, \eta) = F_n(\xi', \eta') = p(t, \sqrt{1 - t^2})\,. \tag{4.13}$$

We can impose another rotation of coordinates, again without changing F_n or $\langle \xi, \eta \rangle$. Let $\tilde{R}$ be the transformation that rotates vectors by

π radians about the x_1-axis, i.e.,

$$\xi' \overset{\tilde{R}}{\longmapsto} \xi'' = (t, -\sqrt{1-t^2}, 0, \ldots, 0),$$

$$\eta' \overset{\tilde{R}}{\longmapsto} \eta'' = (1, 0, \ldots, 0).$$

Just as in (4.13), we conclude that

$$F_n(\xi, \eta) = F_n(\xi'', \eta'') = p(t, -\sqrt{1-t^2}). \tag{4.14}$$

From (4.13) and (4.14), we see that $\sqrt{1-t^2}$ must appear only with even powers in F_n. Thus, F_n is really a polynomial in t and $1-t^2$, which is just a polynomial in t. Since $t = \langle \xi, \eta \rangle$, the lemma is proved. ∎

4.3   Legendre Polynomials

Theorem 4.10 *Let* $\eta = (1, 0, \ldots, 0)$ *and let* $L_n(x)$ *be a harmonic homogeneous polynomial of degree n satisfying*

(i) $L_n(\eta) = 1$,

(ii) $L_n(Rx) = L_n(x)$ *for all rotation matrices R such that $R\eta = \eta$.*

Then $L_n(x)$ is the only harmonic homogeneous polynomial of degree n obeying these properties. In particular, these two properties uniquely determine the corresponding spherical harmonic $L_n|_{S^{p-1}}$. Moreover, this spherical harmonic $L_n(\xi)$ is a polynomial in $\langle \xi, \eta \rangle$.

Proof Let $\xi \in S^{p-1}$. Then there exist $\nu \in S^{p-1}$ such that $\langle \nu, \eta \rangle = 0$ and $t \in [-1, 1]$ such that $\xi = t\eta + \sqrt{1-t^2}\,\nu$. Notice $\langle \xi, \eta \rangle = t$ and that

$$\xi_1 = t, \quad \xi_2^2 + \cdots + \xi_p^2 = 1 - t^2. \tag{4.15}$$

As in the proof of Theorem 4.4, we can write

$$L_n(x) = \sum_{j=1}^{n} x_1^j h_{n-j}(x_2, x_3, \ldots, x_p), \tag{4.16}$$

where the h_{n-j} are homogeneous polynomials of degree $n - j$. Let R be a rotation matrix as described in property (ii) in the statement of the theorem. When R acts on x, it sends

$$x_1, x_2, \ldots, x_p \overset{R}{\longmapsto} x_1, x_2', \ldots, x_p' .$$

By property (ii),

$$0 = L_n(x) - L_n(Rx) = \sum_{j=1}^{n} x_1^j \left[h_{n-j}(x_2, \ldots, x_p) - h_{n-j}(x_2', \ldots, x_p') \right] ,$$

and using the linear independence of the x_1^j, we see that all the $h_{n-j}(x_2, \ldots, x_p)$ are invariant under the rotation of coordinates R. Thus, these polynomials must depend only on the radius $(x_2^2 + \cdots + x_p^2)^{1/2}$, i.e.,

$$h_{n-j}(x_2, \ldots, x_p) = c_{n-j} \left(\sqrt{x_2^2 + \cdots + x_p^2} \right)^{n-j} , \qquad (4.17)$$

where the c_{n-j} are constants, and $c_{n-j} = 0$ for all odd $n - j$ since the h_{n-j} must be polynomials. Property (i) gives us one of these coefficients,

$$1 = L_n(\eta) = c_0$$

since all the $x_2, \ldots, x_p$ are zero for this vector. Since the form of the h_{n-j} is given by (4.17), knowing c_0 is enough to determine the rest of the h_{n-j} using the recursive relation (4.6) and the fact that the c_{n-j} are zero for odd $n - j$. Therefore, L_n as well as the spherical harmonic $L_n|_{S^{p-1}}$ are uniquely determined by properties (i) and (ii). Finally, using (4.16), (4.17), and (4.15),

$$L_n(\xi) = \sum_{\substack{j=0 \\ n-j=\text{even}}}^{n} \xi_1^j c_{n-j} (\xi_2^2 + \cdots \xi_p^2)^{(n-j)/2} ,$$

or

$$L_n(\xi(t)) = \sum_{\substack{j=0 \\ n-j=\text{even}}}^{n} t^j c_{n-j} (1 - t^2)^{(n-j)/2} . \qquad (4.18)$$

That is, $L_n(\xi)$ is a polynomial in $t = \langle \xi, \eta \rangle$, and the theorem is proved. ∎

Definition We call the polynomial $L_n(\xi)$ introduced in Theorem 4.10 the *Legendre polynomial*[2] *of degree* n. Written in terms of the variable t, we denote it by $P_n(t)$.

Remark We can quickly obtain some properties of Legendre polynomials. First, from (4.18), we can see $P_n(t)$ is a polynomial of degree n. We can also compute

$$1 = L_n(\eta) = P_n(\langle \eta, \eta \rangle) = P_n(1).$$

We can determine the parity of the Legendre polynomials in more than one way. First, from (4.18), we can see that if n is even, P_n contains only even powers of t, while if n is odd, P_n contains only odd powers of t. Alternatively, we can see that

$$P_n(-t) = P_n(\langle -\xi, \eta \rangle) = L_n(-\xi) = (-1)^n L_n(\xi)$$
$$= (-1)^n P_n(\langle \xi, \eta \rangle) = (-1)^n P_n(t). \tag{4.19}$$

Either way, we determine that P_n is even whenever n is even, and P_n is odd whenever n is odd.

In the following theorem, we demonstrate how to write the Legendre polynomials in terms of an orthonormal set of spherical harmonics.

Theorem 4.11 (Legendre Polynomials Addition Theorem)
Let $\{Y_{n,j}(\xi)\}_{j=1}^{N(p,n)}$ be an orthonormal set of n-th degree spherical harmonics. Then the Legendre polynomial of degree n may be written as

$$P_n(\langle \xi, \eta \rangle) = \frac{\Omega_{p-1}}{N(p,n)} \sum_{j=1}^{N(p,n)} Y_{n,j}(\xi) Y_{n,j}(\eta). \tag{4.20}$$

---

[2]We follow the naming convention used in [9] and common in physics. Mathematicians usually refer to these as Legendre polynomials only when $p = 3$ and as *ultraspherical polynomials* for arbitrary p.

Proof Since (4.20) is invariant under coordinate rotations, we can choose $\eta = (1, 0, \ldots, 0)$. Consider again the function $F_n(\xi, \eta)$ defined in (4.12). Since we have already defined the vector η, we will think of it as a fixed parameter in the function F_n. For now, we will write $F_n(\xi; \eta)$ and think of F_n as a function of one vector ξ.

First, notice that F_n, being a linear combination of spherical harmonics $Y_{n,j}(\xi)$ (since we consider the $Y_{n,j}(\eta)$ to be constants), is itself a spherical harmonic, i.e., the restriction of a harmonic homogeneous polynomial to the unit sphere. Next, notice that any rotation of coordinates R that leaves η fixed leaves the function F_n invariant. Indeed, using Lemma 4.8 or Lemma 4.9,

$$F(R\xi; \eta) \;=\; F(R\xi; R\eta) \;=\; F(\xi; \eta), \quad \text{for all } R \text{ such that } R\eta = \eta.$$

Notice that in the application of the rotation R, we only rotate the coordinates of ξ in F_n since we consider η a fixed parameter, though this made no difference in the calculation.

Let us normalize the function F_n by dividing it by the constant $F_n(\eta; \eta)$. Thus, we have found a function $F_n(\xi; \eta)/F_n(\eta; \eta)$ that obeys all the properties of the Legendre polynomial described in the statement of Theorem 4.10. Since, by the same theorem, these properties uniquely define the Legendre polynomial, we conclude

$$P_n(\langle \xi, \eta \rangle) = \frac{F_n(\xi, \eta)}{F_n(\eta, \eta)}.$$

To complete the proof, we will compute $F_n(\eta, \eta)$. Since, by Lemma 4.9, $F_n(\eta, \eta)$ depends only on the inner product $\langle \eta, \eta \rangle = 1$, it is a constant. Thus,

$$F_n(\eta, \eta)\Omega_{p-1} = \int_{\eta \in S^{p-1}} F_n(\eta, \eta) \, d\Omega_{p-1}$$

$$= \int_{\eta \in S^{p-1}} \sum_{j=1}^{N(p,n)} Y_{n,j}(\eta)^2 \, d\Omega_{p-1}$$

$$= \sum_{j=1}^{N(p,n)} \int_{\eta \in S^{p-1}} Y_{n,j}(\eta)^2 \, d\Omega_{p-1} = N(p, n),$$

using the orthonormality of the n-th degree spherical harmonics. We see that

$$P_n(\langle \xi, \eta \rangle) = \frac{F_n(\xi, \eta)}{F_n(\eta, \eta)} = \frac{\Omega_{p-1}}{N(p, n)} \sum_{j=1}^{N(p,n)} Y_{n,j}(\xi) \, Y_{n,j}(\eta),$$

and we are finished.   ∎

Obviously the preceding theorem generalizes the well known addition theorem in the three-dimensional space which the reader may have encountered in various settings — perhaps in a course of electromagnetism or a course of mathematical methods in physics among other possibilities. It is probably a worthwhile exercise to reproduce the three-dimensional case so our reader makes the connections. We thus do so immediately.

Example For $p = 3$, let

$$\xi = (\sin\theta \, \cos\phi, \ \sin\theta \, \sin\phi, \ \cos\theta),$$
$$\eta = (\sin\theta' \, \cos\phi', \ \sin\theta' \, \sin\phi', \ \cos\theta').$$

Hence

$$\langle \xi, \eta \rangle = \cos\theta \, \cos\theta' + \sin\theta \, \sin\theta' \, \cos(\phi - \phi').$$

At the same time, if γ indicates the angle formed by the two vectors ξ and η,

$$\langle \xi, \eta \rangle = \cos\gamma.$$

For $p = 3$, we have $N(3,n) = 2n + 1$. To define the orthogonal functions, we first introduce the *associated Legendre polynomials*[3] $P_n^j(t)$ with $j = -n, \ldots, -1, 0, 1, \ldots, n$ by the relations: $P_n^0(t) = P_n(t)$,

$$P_n^j(t) = \frac{(-1)^j}{2^n\, n!} (1 - t^2)^{j/2} \frac{d^{n+j}}{dt^{n+j}} (t^2 - 1)^n, \quad j = 1, 2, \ldots, n,$$

and

$$P_n^{-j}(t) = (-1)^j \frac{(n-j)!}{(n+j)!} P_n^j(t), \quad j = 1, 2, \ldots, n.$$

Then[4]

$$Y_{n,j}(\xi) = \sqrt{\frac{2n+1}{2\pi} \frac{(n-j)!}{(n+j)!}}\, P_n^j(\cos\theta)\, \cos(j\phi), \quad j = 1, 2, \ldots, n,$$

$$\text{(4.21a)}$$

$$Y_{n,j}(\xi) = \sqrt{\frac{2n+1}{2\pi} \frac{(n-j)!}{(n+j)!}}\, P_n^j(\cos\theta)\, \sin(j\phi), \quad j = -1, -2, \ldots, -n,$$

$$\text{(4.21b)}$$

and

$$Y_{n,0}(\xi) = \sqrt{\frac{2n+1}{4\pi}}\, P_n(\cos\theta). \qquad \text{(4.21c)}$$

Notice that, since we are working with real functions, the $Y_{n,j}(\xi)$ with $j \neq 0$ are not the traditional spherical harmonics which are complex:

$$\tilde{Y}_{n,j}(\xi) = \sqrt{\frac{2n+1}{4\pi} \frac{(n-j)!}{(n+j)!}}\, P_n^j(\cos\theta)\, e^{ij\phi},$$

---

[3]Which are not polynomials if j is odd: Notice the fractional exponent $j/2$ of $1 - t^2$.

[4]The inquisitive reader may ask here: "How is it known that the 3-dimensional spherical harmonics can be written in this way?" The answer is given by Problem 6 at the end of chapter. In one sentence: Any spherical harmonic can be expressed as a product of associated Legendre polynomials.

where we used the tilde to differentiate the two conventions. In fact, our real functions are known in the literature as *surface harmonics of the first kind*, and they are furthermore divided into *zonal* (when $j = 0$), *tesseral*[5] (when $j \neq 0, n$) and *sectoral* (when $j = n$); see Figure 4.1. Returning to equation (4.20), since $\Omega_2 = 4\pi$, the addition

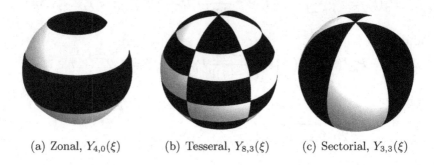

(a) Zonal, $Y_{4,0}(\xi)$ (b) Tesseral, $Y_{8,3}(\xi)$ (c) Sectorial, $Y_{3,3}(\xi)$

Figure 4.1: Graphical representation of three surface harmonics which provides an explanation of the terminology used: the zero lines divide the surface of the sphere in areas that look ribbons, a chessboard, or slices of an orange.

theorem in three dimensions is

$$
\begin{aligned}
P_n(\cos\gamma) &= P_n(\cos\theta)\, P_n(\cos\theta') \\
&+ 2\sum_{j=1}^{n} \frac{(n-j)!}{(n+j)!}\, P_n^j(\cos\theta)\, P_n^j(\cos\theta')\, \cos\big(j(\phi - \phi')\big),
\end{aligned}
$$

which you may have encountered in this form previously (see, for example, equation (3.68) of [10]). However its most popular form is

$$
P_n(\cos\gamma) = \frac{4\pi}{2n+1} \sum_{j=-n}^{n} \tilde{Y}_{nj}(\xi)\, \tilde{Y}_{nj}^*(\xi'),
$$

which can be found in any book discussing spherical harmonics in three dimensions. ∎

---

[5]From the Greek word τέσσερα meaning *four* which, in Latin, became *tesserae* meaning square.

Obviously, there is an addition theorem for the Legendre polynomials in two dimensions although it is not encountered often in books. The following example shows that the addition theorem for $p = 2$ reduces to the functional form of the Legendre polynomials.

Example For $p = 2$, let

$$\xi = (\cos\theta,\ \sin\theta),$$
$$\eta = (\cos\theta',\ \sin\theta').$$

Hence

$$t = \langle\xi,\eta\rangle = \cos(\theta - \theta').$$

Since $N(2, n) = 2$, we choose the orthogonal functions to be[6]

$$Y_{n,1}(\xi) = \frac{1}{\sqrt{\pi}}\cos(n\theta),$$
$$Y_{n,2}(\xi) = \frac{1}{\sqrt{\pi}}\sin(n\theta).$$

Also $\Omega_1 = 2\pi$. Then equation (4.20) takes the form

$$P_n(t) = \cos\left(n(\theta - \theta')\right) = \cos\left(n\cos^{-1}t\right).$$

That is, we have found the functional expression of the Legendre polynomials in two dimensions. In particular, the reader may recognize the expression as that of the *Chebyshev polynomials of the first kind*.   ■

Of course, besides expressing the Legendre polynomials in terms of the spherical harmonics, we can expand the spherical harmonics in terms of Legendre polynomials. To show this, we will need first the following result.

---

[6]See also equation (1.9).

Lemma 4.12 *For any set $\{Y_{n,j}\}_{j=1}^k$ of $k \leq N(p,n)$ linearly independent n-th degree spherical harmonics, there exists a set $\{\eta_i\}_{i=1}^k$ of unit vectors such that the $k \times k$ determinant*

$$
\begin{vmatrix}
Y_{n,1}(\eta_1) & Y_{n,1}(\eta_2) & \cdots & Y_{n,1}(\eta_k) \\
Y_{n,2}(\eta_1) & Y_{n,2}(\eta_2) & \cdots & Y_{n,2}(\eta_k) \\
\vdots & \vdots & \ddots & \vdots \\
Y_{n,k}(\eta_1) & Y_{n,k}(\eta_2) & \cdots & Y_{n,k}(\eta_k)
\end{vmatrix}
\tag{4.22}
$$

is nonzero.

Proof We will prove the lemma by induction on k. First, consider a linearly independent set $\{Y_{n,1}\}$ of one spherical harmonic of degree n. $Y_{n,1}$ cannot be the zero function. Then, there exists a unit vector, call it η_1, such that the determinant $|Y_{n,1}(\eta_1)| = Y_{n,1}(\eta_1) \neq 0$. Thus, the lemma holds for the case $k = 1$. Now, suppose the lemma holds for some $k = \ell - 1 \leq N(p,n) - 1$, and let $\{Y_{n,j}\}_{j=1}^\ell$ be a set of $\ell \leq N(p,n)$ linearly independent n-th degree spherical harmonics. By the induction hypothesis, we can choose a set $\{\eta_i\}_{i=1}^{\ell-1}$ of unit vectors such that the $(\ell - 1) \times (\ell - 1)$ determinant

$$
\Delta_\ell =
\begin{vmatrix}
Y_{n,1}(\eta_1) & Y_{n,1}(\eta_2) & \cdots & Y_{n,1}(\eta_{\ell-1}) \\
Y_{n,2}(\eta_1) & Y_{n,2}(\eta_2) & \cdots & Y_{n,2}(\eta_{\ell-1}) \\
\vdots & \vdots & \ddots & \vdots \\
Y_{n,\ell-1}(\eta_1) & Y_{n,\ell-1}(\eta_2) & \cdots & Y_{n,\ell-1}(\eta_{\ell-1})
\end{vmatrix}
\neq 0.
\tag{4.23}
$$

Now consider the spherical harmonic defined by the $\ell \times \ell$ determinant,

$$
\Delta =
\begin{vmatrix}
Y_{n,1}(\eta_1) & Y_{n,1}(\eta_2) & \cdots & Y_{n,1}(\eta_{\ell-1}) & Y_{n,1}(\xi) \\
Y_{n,2}(\eta_1) & Y_{n,2}(\eta_2) & \cdots & Y_{n,2}(\eta_{\ell-1}) & Y_{n,2}(\xi) \\
\vdots & \vdots & \ddots & \vdots & \vdots \\
Y_{n,\ell-1}(\eta_1) & Y_{n,\ell-1}(\eta_2) & \cdots & Y_{n,\ell-1}(\eta_{\ell-1}) & Y_{n,\ell-1}(\xi) \\
Y_{n,\ell}(\eta_1) & Y_{n,\ell}(\eta_2) & \cdots & Y_{n,\ell}(\eta_{\ell-1}) & Y_{n,\ell}(\xi)
\end{vmatrix}.
\tag{4.24}
$$

If we compute this determinant by performing a cofactor expansion down the last column and indicating by Δ_j the minor determinant corresponding to $Y_{n,j}(\xi)$,

$$\Delta = \sum_{j=1}^{\ell} (-1)^{\ell+j} \Delta_j Y_{n,j}(\xi),$$

we can see that we will have a linear combination of spherical harmonics $Y_{n,j}(\xi)$ which are linearly independent. Thus, for the determinant to vanish identically, all the coefficients of the $Y_{n,j}(\xi)$ must vanish. But notice that the coefficient of $Y_{n,\ell}(\xi)$ is the determinant (4.23) which does not vanish. Thus, the spherical harmonic given by the determinant (4.24) is not the zero function; i.e., there exists a unit vector $\xi = \eta_\ell$ such that the determinant (4.24) is nonzero. Therefore, the lemma holds for the case $k = \ell$ and, by induction, for all $k \leq N(p,n)$. ∎

Theorem 4.13 *For any spherical harmonic $Y_n(\xi)$ of degree n, there exist coefficients a_k and unit vectors η_k such that*

$$Y_n(\xi) = \sum_{k=1}^{N(p,n)} a_k P_n(\langle \xi, \eta_k \rangle).$$

Proof Let $\{Y_{n,j}(\xi)\}_{j=1}^{N(p,n)}$ be an orthonormal set of n-th degree spherical harmonics, and let $Y_n(\xi)$ be any spherical harmonic of degree n. For a unit vector η, we can write

$$P_n(\langle \xi, \eta \rangle) = \frac{\Omega_{p-1}}{N(p,n)} \sum_{j=1}^{N(p,n)} Y_{n,j}(\xi) Y_{n,j}(\eta), \qquad (4.25)$$

by Theorem 4.11. Let us choose a set of unit vectors η_j such that the determinant (4.22) with $k = N(p,n)$ is nonzero; this is possible by Lemma 4.12. Replacing η by η_j in (4.25) for $1 \leq j \leq N(p,n)$ creates the system of equations in which

$$\frac{N(p,n)}{\Omega_{p-1}} \begin{bmatrix} P_n(\langle \xi, \eta_1 \rangle) \\ P_n(\langle \xi, \eta_2 \rangle) \\ \vdots \\ P_n(\langle \xi, \eta_{N(p,n)} \rangle) \end{bmatrix}$$

equals

$$
\begin{bmatrix}
Y_{n,1}(\eta_1) & \cdots & Y_{n,N(p,n)}(\eta_1) \\
Y_{n,1}(\eta_2) & \cdots & Y_{n,N(p,n)}(\eta_2) \\
\vdots & \ddots & \vdots \\
Y_{n,1}(\eta_{N(p,n)}) & \cdots & Y_{n,N(p,n)}(\eta_{N(p,n)})
\end{bmatrix}
\begin{bmatrix}
Y_{n,1}(\xi) \\
Y_{n,2}(\xi) \\
\vdots \\
Y_{n,N(p,n)}(\xi)
\end{bmatrix} .
$$

The determinant of the $N(p,n) \times N(p,n)$ coefficient matrix on the right-hand side of this system has nonzero determinant by our choice of the vectors η_j. Thus, this system is invertible; i.e., there exist coefficients $c_{\ell,m}$ such that

$$
Y_{n,\ell}(\xi) = \sum_{m=1}^{N(p,n)} c_{\ell,m} P_n(\langle \xi, \eta_m \rangle), \quad \text{for each } 1 \le \ell \le N(p,n). \quad (4.26)
$$

Now since any spherical harmonic, and in particular $Y_n(\xi)$, can be expanded in terms of the basis functions $\{Y_{n,i}(\xi)\}_{i=1}^{N(p,n)}$, there exist coefficients a_k such that

$$
Y_n(\xi) = \sum_{k=1}^{N(p,n)} a_k P_n(\langle \xi, \eta_k \rangle),
$$

by (4.26), and the theorem is proved.   ∎

We will now list and prove several basic properties of spherical harmonics and Legendre polynomials, some of which are useful for making estimates. The proofs are straightforward.

Lemma 4.14 *For any spherical harmonic $Y_n(\xi)$,*

$$
Y_n(\xi) = \frac{N(p,n)}{\Omega_{p-1}} \int_{\eta \in S^{p-1}} Y_n(\eta)\, P_n(\langle \xi, \eta \rangle)\, d\Omega_{p-1}. \quad (4.27)
$$

Proof Let $Y_n(\xi)$ be any n-th degree spherical harmonic, and let $\{Y_{n,j}\}_{j=1}^{N(p,n)}$ be an orthonormal set of such functions. Then, we can expand $Y_n(\eta)$ in terms of this basis; i.e., for some coefficients a_j,

$$Y_n(\eta) = \sum_{j=1}^{N(p,n)} a_j \, Y_{n,j}(\eta) \, .$$

Using this expansion and Theorem 4.11 to rewrite $P_n(\langle \xi, \eta \rangle)$, the right-hand side of (4.27) becomes

$$\frac{N(p,n)}{\Omega_{p-1}} \int_{\eta \in S^{p-1}} \left[\sum_{j=1}^{N(p,n)} a_j \, Y_{n,j}(\eta) \right] \left[\frac{\Omega_{p-1}}{N(p,n)} \sum_{k=1}^{N(p,n)} Y_{n,k}(\xi) \, Y_{n,k}(\eta) \right] d\Omega_{p-1} \, ,$$

or

$$\sum_{j,k=1}^{N(p,n)} a_j \, Y_{n,k}(\xi) \left[\int_{\eta \in S^{p-1}} Y_{n,j}(\eta) \, Y_{n,k}(\eta) \, d\Omega_{p-1} \right] = \sum_{j=1}^{N(p,n)} a_j \, Y_{n,j}(\xi)$$

$$= Y_n(\xi) \, ,$$

as required.   ∎

Proposition 4.15 *The Legendre polynomials $P_n(t)$ are bounded*

$$|P_n(t)| \leq 1 \, , \ \textit{for all } t \in [0,1] \tag{4.28}$$

and obey the following normalization condition

$$\int_{\xi \in S^{p-1}} P_n(\langle \xi, \eta \rangle)^2 \, d\Omega_{p-1} = \frac{\Omega_{p-1}}{N(p,n)} \, . \tag{4.29}$$

Proof We rewrite $P_n(t)^2$ using Theorem 4.11. Then, we use the Cauchy-Schwarz inequality, viewing the sum of products in this expression as a dot product, to derive the required result:

$$P_n(\langle \xi, \eta \rangle)^2 = \left[\frac{\Omega_{p-1}}{N(p,n)} \sum_{j=1}^{N(p,n)} Y_{n,j}(\xi) \, Y_{n,j}(\eta) \right]^2$$

$$\leq \left[\frac{\Omega_{p-1}}{N(p,n)} \sum_{j=1}^{N(p,n)} Y_{n,j}(\xi)^2 \right] \left[\frac{\Omega_{p-1}}{N(p,n)} \sum_{j=1}^{N(p,n)} Y_{n,j}(\eta)^2 \right],$$

so

$$P_n(\langle \xi, \eta \rangle)^2 \leq P_n(\langle \xi, \xi \rangle) \, P_n(\langle \eta, \eta \rangle) = P_n(1)^2 = 1,$$

proving the first result.

Again, we use Theorem 4.11 to rewrite the integral on the left side of (4.29) as

$$\int_{\xi \in S^{p-1}} \frac{\Omega_{p-1}^2}{N(p,n)^2} \sum_{j,k=1}^{N(p,n)} Y_{n,j}(\xi) \, Y_{n,j}(\eta) Y_{n,k}(\xi) \, Y_{n,k}(\eta) \, d\Omega_{p-1}$$

which then becomes

$$\frac{\Omega_{p-1}^2}{N(p,n)^2} \sum_{j=1}^{N(p,n)} Y_{n,j}(\eta) \, Y_{n,j}(\eta) = \frac{\Omega_{p-1}}{N(p,n)} P_n(\langle \eta, \eta \rangle) = \frac{\Omega_{p-1}}{N(p,n)},$$

thus proving the second result.   ∎

Proposition 4.16 *The spherical harmonics $Y_n(\xi)$ satisfy the following inequality*

$$|Y_n(\xi)| \leq \sqrt{\frac{N(p,n)}{\Omega_{p-1}} \int_{\eta \in S^{p-1}} Y_n(\eta)^2 \, d\Omega_{p-1}}. \qquad (4.30)$$

Proof We start by taking the square of equation (4.27):

$$Y_n(\xi)^2 = \frac{N(n,p)^2}{\Omega_{p-1}^2} \left[\int_{\eta \in S^{p-1}} Y_n(\eta) \, P_n(\langle \xi, \eta \rangle) \, d\Omega_{p-1} \right]^2.$$

Viewing the integral as an inner product, we apply the Cauchy-Schwarz inequality,

$$Y_n(\xi)^2 \leq \frac{N(n,p)^2}{\Omega_{p-1}^2} \left[\int\limits_{\eta \in S^{p-1}} Y_n(\eta)^2 \, d\Omega_{p-1} \right] \left[\int\limits_{\eta \in S^{p-1}} P_n(\langle \xi, \eta \rangle)^2 \, d\Omega_{p-1} \right].$$

Thus,

$$Y_n(\xi)^2 \leq \frac{N(p,n)}{\Omega_{p-1}} \int\limits_{S^{p-1}} Y_n(\eta)^2 \, d\Omega_{p-1},$$

where we used property (4.29) in the last step.   ∎

We will now begin to investigate the properties of Legendre polynomials as orthogonal polynomials. Let us rewrite the integral in (4.29). Note that the integrand depends only on the inner product $\langle \xi, \eta \rangle$ and that we integrate ξ over the surface of the $(p-1)$-sphere. We can take advantage of these observations to reduce the $(p-1)$-dimensional integral to a one-dimensional integral.

Since we integrate over the entire sphere, we can perform any rotation of coordinates without changing the value of the integral. Let us impose a coordinate change R that aligns the unit vector η along the x_p-axis, which we will picture pointing "north," i.e., take $\eta = (0, \ldots, 0, 1)$. If we let $t = \langle \xi, \eta \rangle$, we can write the unit vector ξ as

$$\xi = \langle \xi, \eta \rangle \eta + (\xi - \langle \xi, \eta \rangle \eta) = t\eta + \sqrt{1 - t^2} \, \nu,$$

for some unit vector ν normal to η. Notice that any such ν gives the same value for the inner product $\langle \xi, \eta \rangle$ and thus the same value for the integrand in (4.29). Note further that the collection of all such vectors ν,

$$\{\nu \in \mathbb{E}^p : |\nu| = 1, \langle \nu, \eta \rangle = 0\},$$

forms a parallel of the $(p-1)$-sphere which is a $(p-2)$-sphere:

$$\{\nu \in \mathbb{E}^p : |\nu| = 1, \nu_p = 0\} = S^{p-2}.$$

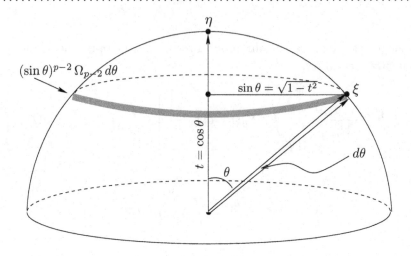

Figure 4.2: In the reduction of a spherically symmetric $(p-1)$-dimensional integral to a 1-dimensional integral, we imagine the $(p-1)$-dimensional sphere as a sum of infinitesimal $(p-2)$-dimensional spheres (parallels) of thickness $d\theta$.

A visual representation is depicted in Figure 4.2.

To get some intuition, let's think of the familiar 3-dimensional case. The 2-dimensional sphere can be thought as a sum of infinitesimal rings oriented along the parallels of the sphere. The infinitesimal ring defined by the azimuthal angle θ has an infinitesimal thickness $d\theta$ and, therefore, it corresponds to a solid angle

$$d\Omega_2 = (2\pi \sin\theta)\, d\theta .$$

The term inside the parenthesis is the length of the ring; multiplied by its thickness it gives the "area" of the ring. We can rewrite $d\Omega_2$ as $-2\pi d(\cos\theta)$ and thus

$$d\Omega_2 = -\Omega_1\, dt$$

in terms of the variable t.

Similarly, in p dimensions

$$d\Omega_{p-1} = \Omega_{p-2} \, (\sin\theta)^{p-2}\, d\theta .$$

The factor $(\sin\theta)^{p-2}$ is easy to explain: The radius of the $(p-2)$-sphere is $R = \sin\theta$; its volume will be proportional to R^{p-2}. It is straightforward to express $d\Omega_{p-1}$ in terms of t:

$$d\Omega_{p-1} = -\Omega_{p-2}(\sin\theta)^{p-3}\, d(\cos\theta)$$
$$= -\Omega_{p-2}(1-t^2)^{\frac{p-3}{2}}\, dt\,.$$

Now we are ready to write (4.29) as a 1-dimensional integral over t. Indeed,

$$\int_{\xi\in S^{p-1}} P_n(\langle\xi,\eta\rangle)^2\, d\Omega_{p-1} = \int_{-1}^{1} P_n(t)^2\,(1-t^2)^{\frac{p-3}{2}}\Omega_{p-2}\, dt\,,$$

which implies

$$\int_{-1}^{1} P_n(t)^2\,(1-t^2)^{\frac{p-3}{2}}\, dt = \frac{\Omega_{p-1}}{N(p,n)\Omega_{p-2}}\,.$$

This expression gives us the norm of the n-th Legendre polynomial with respect to the weight $w(t) = (1-t^2)^{\frac{p-3}{2}}$, namely

$$\|P_n(t)\|_w = \sqrt{\langle P_n(t), P_n(t)\rangle_w} = \sqrt{\frac{\Omega_{p-1}}{N(p,n)\,\Omega_{p-2}}}\,. \qquad (4.31)$$

Note that in coming up with this fact, we have essentially proved the following lemma, which will be used to show the next few results.

Lemma 4.17 *Let η be a unit vector and f be a function. Then*

$$\int_{\xi\in S^{p-1}} f(\langle\xi,\eta\rangle)\, d\Omega_{p-1} = \Omega_{p-2}\int_{-1}^{1} f(t)(1-t^2)^{(p-3)/2}\, dt\,.$$

Theorem 4.18 *Any two distinct Legendre polynomials $P_n(t), P_m(t)$ are orthogonal over the interval $[-1, 1]$ with respect to the weight $(1 - t^2)^{\frac{p-3}{2}}$. That is,*

$$\int_{-1}^{1} P_n(t) P_m(t) (1 - t^2)^{\frac{p-3}{2}} \, dt = 0, \quad for \ n \neq m.$$

Proof Let $\eta = (1, 0, \ldots, 0)$. The Legendre polynomial $P_n(\langle \xi, \eta \rangle)$ is equal to the spherical harmonic $L_n(\xi)$ having the properties listed in Theorem 4.10. By Theorem 4.6,

$$0 = \int_{\xi \in S^{p-1}} L_n(\xi) L_m(\xi) \, d\Omega_{p-1} = \int_{\xi \in S^{p-1}} P_n(\langle \xi, \eta \rangle) P_m(\langle \xi, \eta \rangle) \, d\Omega_{p-1},$$

for $n \neq m$. Since the integral on the right-hand side of the above equation depends only on the inner product $\langle \xi, \eta \rangle$, we can use Lemma 4.17 to rewrite this integral as

$$\Omega_{p-2} \int_{-1}^{1} P_n(t) P_m(t) (1 - t^2)^{\frac{p-3}{2}} \, dt = 0, \quad for \ n \neq m,$$

thus completing the proof.   ∎

The above theorem allows us to use the results from Chapter 3 to write down more properties of the Legendre polynomials. Using the comment just above Section 3.4, we can find the Rodrigues formula for the Legendre polynomials in p dimensions. Theorem 4.18 tells us that the Legendre polynomials are orthogonal with respect to $w(t) = (1 - t^2)^{(p-3)/2}$ so that

$$P_n(t) = c_n (1 - t^2)^{-(p-3)/2} \left(\frac{d}{dt} \right)^n \left[(1 - t^2)^{(p-3)/2} (1 - t^2)^n \right]$$

$$= c_n (1 - t^2)^{(3-p)/2} \left(\frac{d}{dt} \right)^n (1 - t^2)^{n+(p-3)/2},$$

and it remains to compute c_n. We know that $P_n(1) = 1$, so

$$1 = c_n(1 - t^2)^{(3-p)/2} \left(\frac{d}{dt} \right)^n (1 - t^2)^{n+(p-3)/2} \Big|_{t=1}.$$

Carrying out two differentiations,

$$1 = c_n(1 - t^2)^{(3-p)/2} \times$$
$$\left(\frac{d}{dt} \right)^{n-1} \left(\frac{n + (p-3)}{2} \right) (-2t)(1 - t^2)^{n-1+(p-3)/2} \Big|_{t=1}$$
$$= c_n(1 - t^2)^{(3-p)/2} \times$$
$$\left(\frac{d}{dt} \right)^{n-2} \left[\left(\frac{n + (p-3)}{2} \right)_2 (-2t)^2 (1 - t^2)^{n-2+(p-3)/2} + \cdots \right] \Big|_{t=1},$$

where we have used the falling factorial notation and left terms that are higher order in $1 - t^2$ in the "$\cdots$" because they will vanish when we substitute $t = 1$. Continuing the pattern,

$$1 = c_n(1 - t^2)^{(3-p)/2} \times$$
$$\left[(n + (p-3)/2)_n (-2t)^n (1 - t^2)^{n-n+(p-3)/2} + \cdots \right] \Big|_{t=1}$$
$$= c_n (n + (p-3)/2)_n (-2)^n,$$

implying that

$$c_n = \frac{(-1)^n}{2^n (n + (p-3)/2)_n}.$$

We have shown the following.

Proposition 4.19 (Rodrigues Formula)

$$P_n(t) = \frac{(-1)^n}{2^n (n + (p-3)/2)_n} (1 - t^2)^{(3-p)/2} \left(\frac{d}{dt} \right)^n (1 - t^2)^{n+(p-3)/2}.$$
$$(4.32)$$

The above Rodrigues formula allows us to prove the following two properties of the Legendre polynomials.

Proposition 4.20 *In p dimensions, the Legendre polynomial $P_n(t)$ of degree n satisfies the differential equation*

$$(1 - t^2)P_n''(t) + (1 - p)tP_n'(t) + n(n + p - 2)P_n(t) = 0. \qquad (4.33)$$

In what follows, we will often write P_n and w instead of $P_n(t)$ and $w(t)$ for simplicity. A portion of the proof will be left as a straightforward computation for the reader.

Proof Consider the expression $\frac{d}{dt}\left[(1 - t^2)^{(p-1)/2}P_n'\right]$. We can use the product rule to rewrite this as

$$\left(\frac{p - 1}{2}\right)(1 - t^2)^{(p-3)/2}(-2t)P_n' + (1 - t^2)^{(p-1)/2}P_n'',$$

or,

$$w\left[(1 - p)tP_n' + (1 - t^2)P_n''\right]. \qquad (4.34)$$

The term in brackets is a polynomial of degree n which can be written as a linear combination of the first n Legendre polynomials. Hence

$$\frac{d}{dt}\left[(1 - t^2)wP_n'\right] = w\sum_{j=0}^{n} c_j P_j. \qquad (4.35)$$

Multiplying each side by P_k, where $0 \le k \le n$ and integrating over the interval $[-1, 1]$, we get

$$c_k\|P_k\|^2 = \int_{-1}^{1} P_k\frac{d}{dt}\left[(1 - t^2)wP_n'\right]dt.$$

Integrating the right-hand side by parts now twice and noticing that the boundary terms vanish, we find

$$c_k\|P_k\|^2 = \underbrace{P_k(1 - t^2)wP_n'\big|_{-1}^{1}}_{\text{vanishes}} - \int_{-1}^{1} P_k'(1 - t^2)wP_n'\,dt$$

$$= \underbrace{-P_k'(1 - t^2)wP_n\big|_{-1}^{1}}_{\text{vanishes}} + \int_{-1}^{1}\left\{\frac{d}{dt}\left[P_k'(1 - t^2)w\right]\right\}P_n\,dt.$$

The expression in curly brackets can be written as wq_k, where q_k is a polynomial of degree k. We can then use Proposition 3.3 for the two integrals to discover that $c_k = 0$ for all $k < n$. Returning to equation (4.35), this result implies that

$$\frac{d}{dt}\left[(1-t^2)wP_n'\right] = wc_nP_n\,.$$

To determine c_n, we will compute the coefficient of the highest power of t on each side of the above equation, using *Newton's binomial expansion*. By keeping only the highest-order term in each binomial expansion, the reader can show that the left-hand side becomes

$$\frac{(-1)^{(p-1)/2}}{2^n}\,\frac{(2n+p-3)_n}{(n+(p-3)/2)_n}\,n(n+p-2)t^{n+p-3}+\cdots,$$

while the right-hand side becomes

$$c_n\frac{(-1)^{(p-3)/2}}{2^n}\,\frac{(2n+p-3)_n}{(n+(p-3)/2)_n}\,t^{n+p-3}+\cdots$$

Comparing the above equations, we conclude that $c_n = -n(n+p-2)$. Inserting this into (4.35) and using (4.34), we find

$$(1-t^2)P_n'' + (1-p)tP_n' + n(n+p-2)P_n = 0\,,$$

completing the proof. ∎

Proposition 4.21 *The Legendre polynomials in p dimensions satisfy the recurrence relation*

$$(n+p-2)\,P_{n+1}(t) - (2n+p-2)\,t\,P_n(t) + n\,P_{n-1}(t) \;=\; 0\,. \quad (4.36)$$

Just as in the previous proof, we will leave part of the following proof to the reader.

Proof We know from Proposition 3.5 that a relation of the form

$$P_{n+1} - (A_nt + B_n)P_n + C_nP_{n-1} = 0 \qquad (4.37)$$

exists. We can quickly determine B_n. Recall from (4.19) that P_n is even (respectively, odd) whenever n is even (respectively, odd). Rewriting the above equation as

$$P_{n+1} - A_n t P_n + C_n P_{n-1} = B_n P_n \, ,$$

we have an odd polynomial equal to an even polynomial, unless both sides vanish. Since the first option is not possible, we conclude that both sides of the equation must cancel, which implies that $B_n = 0$. We will use the notation and results of Proposition 3.5 to compute A_n, C_n. Keeping only the highest-order terms in each binomial expansion in the Rodrigues formula (4.32) and carrying out the derivatives, it is straightforward to show that the leading coefficient of P_n is

$$k_n = \frac{(2n+p-3)_n}{2^n(n+(p-3)/2)_n} \, ,$$

from which we determine the leading coefficient of P_{n+1},

$$k_{n+1} = \frac{(2n+p-1)_{n+1}}{2^{n+1}(n+(p-1)/2)_{n+1}} \, ,$$

allowing us to compute

$$A_n = \frac{k_{n+1}}{k_n} = \frac{2n+p-2}{n+p-2} \, .$$

Now, using (4.31),

$$
\begin{aligned}
C_n &= \frac{A_n}{A_{n-1}} \frac{\|P_n\|^2}{\|P_{n-1}\|^2} \\
&= \frac{2n+p-2}{n+p-2} \frac{n+p-3}{2n+p-4} \frac{\Omega_{p-1}}{N(p,n)\Omega_{p-2}} \frac{N(p,n-1)\Omega_{p-2}}{\Omega_{p-1}} \, .
\end{aligned}
$$

Using Theorem 4.4, we can compute

$$C_n = \frac{n}{n+p-2} \, .$$

Inserting these results into (4.37) and multiplying by $n+p-2$, we find the required result. ■

It is perhaps known[7] to the reader that, in $p = 3$, the Legendre polynomials are a special case of the Gegenbauer polynomials $C_n^{(\alpha)}(t)$ — namely, $P_n(t) = C_n^{(1/2)}(t)$ — which are themselves a special case of the Jacobi polynomials. In particular, the Gegenbauer polynomials satisfy the recursion relation

$$(n+1)\,C_{n+1}^{(\alpha)}(t) - (2n+2\alpha)\,t\,C_n^{(\alpha)}(t) + (n+2\alpha-1)\,C_{n-1}^{(\alpha)}(t) \;=\; 0\,.$$

For $\alpha = (p-2)/2$

$$(n+1)\,C_{n+1}^{(\frac{p-2}{2})}(t) - (2n+p-2)\,t\,C_n^{(\frac{p-2}{2})}(t) + (n+p-3)\,C_{n-1}^{(\frac{p-2}{2})}(t) \;=\; 0\,. \tag{4.38}$$

For $p = 3$ this equation becomes identical to the one satisfied by $P_n(t)$ in the same dimensions (in accordance with the previous comment). For any other p, the two equations differ in two of their three coefficients. However, one can show that the Legendre polynomials are still related to the Gegenbauer polynomials for any p by a multiplicative factor

$$C_n^{(\frac{p-2}{2})}(t) \;=\; \binom{n+p-3}{n}\,P_n(t)\,. \tag{4.39}$$

It is also interesting to observe that once we find the Legendre polynomials in $p = 2$ and $p = 3$ dimensions, they are determined for all higher dimensions. In particular, the following theorem proves, more specifically, that the Legendre polynomials for even p can be found from the $p = 2$ case and that those for odd p can be found from the $p = 3$ case. For the next theorem, we will let $P_{n,p}(t)$ denote the n-th degree Legendre polynomial in p dimensions.

Theorem 4.22 *For all* $j = 0, 1, \ldots, n$,

$$P_{n-j,2j+p}(t) \;=\; \frac{((p-1)/2)_j}{(-n)_j(n+p-2)_j} \left(\frac{d}{dt}\right)^j P_{n,p}(t)\,.$$

---

[7]The reader will benefit from looking at the tables with the properties of the classical orthogonal polynomials found in [1] as s/he will get a bird's eye view of the patterns and relations.

Proof Differentiating (4.33) once with respect to t, we get

$$(1 - t^2)P'''_{n,p} + (-1 - p)tP''_{n,p} + (n - 1)(n + p - 1)P'_{n,p} = 0,$$

but this is just (4.33) again with the following substitutions:

$$P_{n,p} \longmapsto P'_{n,p},$$
$$p \longmapsto p + 2,$$
$$n \longmapsto n - 1.$$

Thus, solving this new differential equation for $P'_{n,p}$ will give

$$P'_{n,p}(t) \propto P_{n-1,p+2}(t).$$

Continuing in this way, if we differentiate (4.33) j times we see that

$$\left(\frac{d}{dt}\right)^j P_{n,p}(t) \propto P_{n-j,p+2j}(t)$$

for any $0 \leq j \leq n$. Since $P_{n-j,p+2j}(1) = 1$, in order to create an equality we must divide the left side of the above equation by its value at $t = 1$, i.e.,

$$P_{n-j,p+2j}(t) = \frac{\left(\frac{d}{dt}\right)^j P_{n,p}(t)}{\left(\frac{d}{dt}\right)^j P_{n,p}(t)\big|_{t=1}}.$$

Using the Rodrigues formula (4.32),

$$\left(\frac{d}{dt}\right)^j P_{n,p}(t)\big|_{t=1} = \frac{(-n)_j(n + p - 2)_j}{((p - 1)/2)_j},$$

as the reader can verify by computation. Therefore,

$$P_{n-j,2j+p}(t) = \frac{((p - 1)/2)_j}{(-n)_j(n + p - 2)_j}\left(\frac{d}{dt}\right)^j P_{n,p}(t),$$

completing the proof. ∎

The following lemma will be useful in the proof of the theorem that follows immediately afterwards.

Lemma 4.23 *Let ξ and ζ be unit vectors, f a function, and $F(\zeta, \xi)$ given by*

$$F(\zeta, \xi) = \int\limits_{\eta \in S^{p-1}} f(\langle \xi, \eta \rangle) P_n(\langle \eta, \zeta \rangle)\, d\Omega_{p-1}\,.$$

Then,

$$F(\zeta, \xi) = \Omega_{p-2}\, P_n(\langle \xi, \zeta \rangle) \int\limits_{-1}^{1} f(t)\, P_n(t)\, (1 - t^2)^{\frac{p-3}{2}}\, dt\,.$$

Proof First, notice that F is invariant under any coordinate rotation R, i.e., $F(R\zeta, R\xi) = F(\zeta, \xi)$. Indeed, the rotation

$$\xi \xmapsto{R} \xi' = R\xi,$$
$$\zeta \xmapsto{R} \zeta' = R\zeta,$$

can be undone by a change of variables

$$\eta \xmapsto{R} \eta' = R\eta,$$

in the integration. Thus, we are allowed to choose our coordinates such that

$$\xi = (1, 0, \ldots, 0), \quad \zeta = (s, \sqrt{1 - s^2}, 0, \ldots, 0),$$

so that $\langle \zeta, \xi \rangle = s$.

Let us think of F as a function of ζ alone and ξ as a fixed parameter; we will write $F(\zeta; \xi)$. The argument of F, namely ζ, only shows up inside the Legendre polynomial $P_n(\langle \eta, \zeta \rangle)$, which is a spherical harmonic — in particular, a harmonic homogeneous polynomial of degree n in the variables $\zeta_1, \zeta_2, \ldots, \zeta_p$. Therefore, the function F is also a harmonic homogeneous polynomial in the components of ζ, i.e., in s, $\sqrt{1-s^2}$. But, since we equally well could have chosen co-ordinates such that $\zeta = (s, -\sqrt{1-s^2}, 0, \ldots, 0)$, F must really be a polynomial in s.

Then, being only a function of the inner product $s = \langle \zeta, \xi \rangle$, we see that F satisfies all the defining properties of the Legendre polynomials in Theorem 4.10 except (i). We conclude

$$F(\zeta, \xi) = c\, P_n(s), \quad \text{for some constant } c.$$

We can determine the constant by considering the case $s = 1$, i.e., $\zeta = \xi$, where $F(\xi, \xi) = cP_n(1) = c$. Using Lemma 4.17,

$$c = \int_{\eta \in S^{p-1}} f(\langle \xi, \eta \rangle)\, P_n(\langle \eta, \xi \rangle)\, d\Omega_{p-1}$$

$$= \Omega_{p-2} \int_{-1}^{1} f(t)\, P_n(t)\, (1-t^2)^{\frac{p-3}{2}}\, dt,$$

as sought. ∎

Theorem 4.24 (Hecke-Funk Theorem) *Let ξ be a unit vector, f a function, and Y_n a spherical harmonic. Then,*

$$\int_{\eta \in S^{p-1}} f(\langle \xi, \eta \rangle)\, Y_n(\eta)\, d\Omega_{p-1} = \Omega_{p-2} Y_n(\xi) \int_{-1}^{1} f(t)\, P_n(t)\, (1-t)^{\frac{p-3}{2}}\, dt.$$

$$(4.40)$$

Proof By Theorem 4.13, there exist a set of coefficients $\{a_k\}_{k=1}^{N(p,n)}$ and a set of unit vectors $\{\zeta_k\}_{k=1}^{N(p,n)}$ such that

$$Y_n(\eta) = \sum_{k=1}^{N(p,n)} a_k\, P_n(\langle \eta, \zeta_k \rangle).$$

Then, we can rewrite the left-hand side of (4.40) as

$$\sum_{k=1}^{N(p,n)} a_k \int_{\eta \in S^{p-1}} f(\langle \xi, \eta \rangle)\, P_n(\langle \eta, \zeta_k \rangle)\, d\Omega_{p-1},$$

or, using Lemma 4.23,

$$\Omega_{p-2}\left(\sum_{k=1}^{N(p,n)} a_k P_n(\langle \xi, \zeta_k \rangle)\right) \int_{-1}^{1} f(t)\, P_n(t)\, (1-t^2)^{\frac{p-3}{2}}\, dt\,.$$

That is

$$\Omega_{p-2} Y_n(\xi) \int_{-1}^{1} f(t)\, P_n(t)\, (1-t)^{\frac{p-3}{2}}\, dt\,,$$

which is exactly the right-hand side of (4.40).   ∎

We wish to find now an integral representation of the Legendre polynomials. Towards this goal, we first prove a lemma.

Consider a vector η of the $(p-1)$-dimensional sphere S^{p-1}. Without loss of generality, we may take it along the x_1-axis. With $\{\eta\}^{\perp}$ we indicate the set of all vectors that are perpendicular to η; this is obviously a hyperplane. The set $\{\eta\}^{\perp} \cap S^{p-1}$ is then the equator of S^{p-1} (a $(p-2)$-sphere) that is orthogonal to η; we will indicate it by S_{η}^{p-1}.

Lemma 4.25 *Let $\eta = (1, 0, \ldots, 0)$ and $x \in \mathbb{E}^p$. Then the function*

$$L_n(x) \;=\; \frac{1}{\Omega_{p-2}} \int_{\zeta \in S_{\eta}^{p-1}} (\langle x, \eta \rangle + i \langle x, \zeta \rangle)^n\, d\Omega_{p-2}\,, \qquad (4.41)$$

when restricted to the sphere, is the n-th Legendre polynomial: $L_n(\xi) = P_n(\langle \xi, \eta \rangle)$.

Proof Clearly, $L_n(x)$ is a polynomial in the components of x. By the binomial theorem, this polynomial is homogeneous of degree n. We can also see that $L_n(x)$ is harmonic. Indeed, applying the Laplace operator on (4.41) and switching the order of differentiation and integration, the integrand becomes

$$\sum_{j=1}^{p} \frac{\partial^2}{\partial x_j^2} (\langle x, \eta \rangle + i \langle x, \zeta \rangle)^n\,,$$

or

$$\sum_{j=1}^{p} \frac{\partial}{\partial x_j} \left[n \left(\langle x, \eta \rangle + i \langle x, \zeta \rangle \right)^{n-1} (\eta_j + i\zeta_j) \right],$$

or

$$n(n-1) \left(\langle x, \eta \rangle + i \langle x, \zeta \rangle \right)^{n-2} \sum_{j=1}^{p} (\eta_j + i\zeta_j)^2.$$

But

$$\sum_{j=1}^{p} (\eta_j + i\zeta_j)^2 = \langle \eta, \eta \rangle - \langle \zeta, \zeta \rangle + 2i \langle \eta, \zeta \rangle = 0,$$

since $\{\eta, \zeta\}$ is an orthonormal set. Also, notice

$$L_n(\eta) = \frac{1}{\Omega_{p-2}} \int_{\zeta \in S_\eta^{p-1}} \left[\langle \eta, \eta \rangle + i \langle \eta, \zeta \rangle \right]^n \, d\Omega_{p-2}$$

$$= \frac{1}{\Omega_{p-2}} \int_{\zeta \in S_\eta^{p-1}} d\Omega_{p-2} = 1.$$

Now, let R be any coordinate rotation leaving η invariant, i.e., let R be any rotation about the x_1-axis. The integrand of $L_n(Rx)$ is then

$$\left[\langle Rx, \eta \rangle + i \langle Rx, \zeta \rangle \right]^n = \left[\langle x, \eta \rangle + i \langle x, R^t \zeta \rangle \right]^n,$$

which can be reset to

$$\left[\langle x, \eta \rangle + i \langle x, \zeta \rangle \right]^n$$

by a change of variables $\zeta \to \zeta' = R\zeta$ and thus $L_n(x)$ is invariant under all such coordinate rotations.

We have shown that $L_n(x)$ has all the properties described in Theorem 4.10, so that, when restricted to the sphere, it becomes the n-th Legendre polynomial. ∎

Now we can prove the following integral representation for $P_n(t)$.

Theorem 4.26

$$P_n(t) \;=\; \frac{\Omega_{p-3}}{\Omega_{p-2}} \int\limits_{-1}^{1} \left(t + is\sqrt{1-t^2}\right)^n (1-s^2)^{\frac{p-4}{2}}\, ds\,.$$

Proof From Lemma 4.41, we have

$$P_n(\langle \xi, \eta \rangle) \;=\; \frac{1}{\Omega_{p-2}} \int\limits_{\zeta \in S_\eta^{p-1}} \left(\langle \xi, \eta \rangle + i\langle \xi, \zeta \rangle\right)^n d\Omega_{p-2}\,.$$

Choose a constant t and unit vector ν such that $\xi = t\eta + \sqrt{1-t^2}\,\nu$ and $\langle \nu, \eta \rangle = 0$. Then, the above equation becomes, using Lemma 4.17 and replacing p by $p-1$,

$$P_n(t) = \frac{1}{\Omega_{p-2}} \int\limits_{\zeta \in S_\eta^{p-1}} \left(t + i\sqrt{1-t^2}\langle \nu, \zeta \rangle\right)^n d\Omega_{p-2}$$

$$= \frac{\Omega_{p-3}}{\Omega_{p-2}} \int\limits_{-1}^{1} \left(t + is\sqrt{1-t^2}\right)^n (1-s^2)^{\frac{p-4}{2}}\, ds\,,$$

as sought.   ∎

4.4   Boundary Value Problems

We conclude this discussion with an application of the ideas we have developed to boundary value problems, where they display most of their physical importance.

We know from Proposition 3.16 that if a set of functions in a Hilbert space is closed, then it is complete. In the following theorem, we will see that a maximal linearly independent set of spherical harmonics of all degrees is closed and thus complete. This result allows us to develop expansions of functions as linear combinations of spherical harmonics, which will be useful in application to boundary value problems. In what follows, let

$$S = \{Y_{n,j} : n \in \mathbb{N}_0, \ 1 \le j \le N(p,n)\},$$

be a maximal set of orthogonal spherical harmonics, and let $\sum_{n,j}$ denote the sum $\sum_{n=0}^{\infty} \sum_{j=1}^{N(p,n)}$.

Theorem 4.27 *Let the function $f : S^{p-1} \to \mathbb{E}$ be continuous. If f is orthogonal to the set S, i.e., if*

$$\int\limits_{\xi \in S^{p-1}} f(\xi) Y_{n,j}(\xi) \, d\Omega_{p-1} = 0, \ \text{for all } n, j,$$

then f is the zero function, i.e.,

$$f(\xi) = 0, \ \text{for all } \xi \in S^{p-1}.$$

The requirement that f be continuous is actually too strict, and the theorem really applies to all square-integrable functions f, i.e., all f such that

$$\int\limits_{S^{p-1}} f(\xi)^2 \, d\Omega_{p-1} < \infty.$$

However, we will only prove the weaker version of this theorem.

Proof We will prove it by contradiction. Suppose that f satisfies the hypotheses of the above theorem, i.e., is continuous and orthogonal to S, but is not the zero function. Then there exists some $\eta \in S^{p-1}$ such that $f(\eta) \ne 0$. We can assume that $f(\eta) > 0$, for if $f(\eta)$ is negative we could consider $-f$ instead. By the continuity of f, there is some neighborhood around η on the sphere where f is positive. That is, there exists a constant s such that $f(\xi) > 0$ whenever $s \le \langle \xi, \eta \rangle \le 1$. Define the function

$$\psi(t) = \begin{cases} 1 - \dfrac{(1-t)^2}{(1-s)^2} & \text{if } s \le t \le 1, \\ 0 & \text{if } -1 \le t \le s. \end{cases}$$

For $s \leq \langle \xi, \eta \rangle \leq 1$, the product $f(\xi)\psi(\langle \xi, \eta \rangle)$ is positive, and it vanishes for all other ξ. Thus,

$$\int_{\xi \in S^{p-1}} f(\xi)\,\psi(\langle \xi, \eta \rangle)\,d\Omega_{p-1}c > 0. \tag{4.42}$$

For simplicity, we will indicate the above integral by c. By the Weierstrass approximation theorem (Proposition 3.10), we can find a polynomial $p(t)$ for any given $\epsilon > 0$ such that

$$|\psi(t) - p(t)| \leq \epsilon, \text{ for all } t \in [-1, 1].$$

For any such ϵ and $p(t)$,

$$\int_{\xi \in S^{p-1}} f(\xi)\,[\psi(\langle \xi, \eta \rangle) - p(\langle \xi, \eta \rangle)]\,d\Omega_{p-1}$$

$$\leq \left| \int_{\xi \in S^{p-1}} f(\xi)\,[\psi(\langle \xi, \eta \rangle) - p(\langle \xi, \eta \rangle)]\,d\Omega_{p-1} \right|$$

$$\leq \int_{\xi \in S^{p-1}} |f(\xi)|\,|\psi(\langle \xi, \eta \rangle) - p(\langle \xi, \eta \rangle)|\,d\Omega_{p-1}$$

$$\leq \epsilon \int_{\xi \in S^{p-1}} |f(\xi)|\,d\Omega_{p-1}.$$

Since $f(\xi)$ is continuous and $\xi \in S^{p-1}$, there exists an M such that $M \geq |f(\xi)|$ for any $\xi \in S^{p-1}$. This implies

$$\int_{\xi \in S^{p-1}} f(\xi)p(\langle \xi, \eta \rangle)\,d\Omega_{p-1} \geq c - M\epsilon\,\Omega_{p-1}.$$

And since we can choose ϵ arbitrarily small,

$$\int_{\xi \in S^{p-1}} f(\xi)\,p(\langle \xi, \eta \rangle)\,d\Omega_{p-1} > 0. \tag{4.43}$$

For the remainder of the proof, we fix ϵ and the corresponding $p(t)$ for which this expression is true.

Now, let m denote the degree of $p(t)$. We can write the polynomial $p(t)$ as a linear combination of the first m Legendre polynomials since each $P_n(t)$ is of degree n; i.e., we can find c_k such that

$$p(t) = \sum_{k=0}^{m} c_k P_k(t).$$

We can thus rewrite the integral in (4.43) as

$$\int_{\xi \in S^{p-1}} f(\xi) \sum_{k=0}^{m} c_k P_k(\langle \xi, \eta \rangle) \, d\Omega_{p-1}.$$

But since the Legendre polynomials are just a special collection of spherical harmonics in ξ, this integral must vanish by hypothesis, contradicting the assertion in (4.43). Therefore, our initial assumption that f is not the zero function must be false. ∎

So for any reasonable function f defined on the sphere, we can write

$$f(\xi) = \sum_{n,j} c_{n,j} Y_{n,j}(\xi). \tag{4.44}$$

To find $c_{n',j'}$, multiply both sides of this equation by $Y_{n'j'}$ and integrate over the sphere. Using the orthonormality of the set S, we find

$$c_{n',j'} = \int_{S^{p-1}} f(\xi) Y_{n',j'}(\xi) \, d\Omega_{p-1}. \tag{4.45}$$

This expansion will be used in the following demonstration.

Problem Consider the following boundary-value problem. Find V in the closed unit ball $\bar{B}^p$, such that

$$\Delta_p V = 0, \quad \text{and} \quad V = f(\xi), \text{ for all } \xi \in S^{p-1}. \tag{4.46}$$

Solution We know that harmonic homogeneous polynomials are solutions to the Laplace equation, as well as any linear combination of them. We have also seen in (4.7) that we can write each of these polynomials as a power of the radius multiplied by a spherical harmonic. Thus, we can construct the solution to (4.46) as a linear combination of $r^n Y_{n,j}(\xi)$ terms. To satisfy the boundary condition, we can use the coefficients in (4.45) to find

$$
V = \sum_{n,j} r^n c_{n,j} Y_{n,j}(\xi) = \sum_{n,j} r^n Y_{n,j}(\xi) \int_{S^{p-1}} f(\eta) Y_{n,j}(\eta) \, d\Omega_{p-1},
$$

$$(4.47)$$

thus solving the problem with the solution being in the form of a series. ∎

In the next subsection, we will learn to solve this problem by another method. Since the solution of this boundary-value problem is unique (as we know from the theory of differential equations), we can equate the answers. And, amazingly, this procedure will give us a generating function for the Legendre polynomials $P_n(t)$.

Before we end this section, let's add an important comment. In writing down equation (4.47), we have assumed that the function V is regular at $r = 0$. However, if this condition is lifted (for example, if the point $r = 0$ is not included on the domain of V), then the expansion (4.47) is not valid anymore. However, the modified expansion is easy to find. If $\Delta_p V = 0$ we express the Laplacian in the form (2.17) and use separation of variables $V = R(r) Y_n(\xi)$:

$$
\frac{1}{r^{p-1} R} \frac{\partial}{\partial r} \left(r^{p-1} \frac{\partial R}{\partial r} \right) + \frac{1}{r^2 Y_n} \Delta_{S^{p-1}} Y_n = 0.
$$

or

$$
r^2 \frac{\partial^2 R}{\partial r^2} + (p-1) r \frac{\partial R}{\partial r} + n(2-p-n) R = 0.
$$

This equation has Euler's form; so we try solutions $R = r^\ell$:

$$
\ell(\ell - 1) + (p-1)\ell + n(2-p-n) = 0.
$$

The two solutions of this equation are

$$\ell_+ = n\,, \quad \ell_- = 2 - p - n\,.$$

Then the expansion of V takes the form

$$V = \sum_{n,j} \left(c_{n,j}\, r^n + \frac{d_{n,j}}{r^{p+n-2}} \right) Y_{n,j}(\xi)\,, \qquad (4.48)$$

where $c_{n,j}, d_{n,j}$ are constants that can be determined by the boundary conditions.

Green's Functions

Given a differential equation

$$D_x(f(x)) = 0\,,$$

where D_x is a differential operator acting on the unknown function $f(x)$, the corresponding Green's function G is defined by the equation

$$D_x(G(x)) = \delta(x - x_0)\,,$$

where the function δ appearing in the right-hand side is the Dirac delta function defined by

$$\delta(x - x_0) = 0\,, \text{ for all } x \neq x_0\,,$$

and

$$\int_{B_\epsilon^p(x_0)} \delta(x - x_0)\, d^p x = 1\,, \text{ for all } \epsilon > 0\,.$$

Let's assume that the given differential operator is the Laplacian in p dimensions, that is, we seek the Green's function which satisfies

$$\Delta_p \tilde{G} = \delta(x - x_0)\,. \qquad (4.49)$$

If we think of $\tilde{G}$ as electric potential, then (4.49) describes the electric potential caused by a point charge at $x_0 \in \mathbb{E}^p$.

Since the Laplacian is invariant under translations, we see that $\tilde{G}$ can only depend on the distance from x_0, i.e., on $\rho = |x - x_0|$. When $\rho \neq 0$, $\tilde{G}$ must satisfy Laplace's equation:

$$
\begin{aligned}
0 = \Delta_p \tilde{G}(\rho) &= \sum_{i=1}^{p} \frac{\partial^2}{\partial x_i^2} \tilde{G}(\rho) \\
&= \sum_{i=1}^{p} \frac{\partial}{\partial x_i} \left(\frac{\partial \tilde{G}(\rho)}{\partial \rho} \frac{\partial \rho}{\partial x_i} \right) \\
&= \tilde{G}''(\rho) \sum_{i=1}^{p} \left(\frac{\partial \rho}{\partial x_i} \right)^2 + \tilde{G}'(\rho) \sum_{i=1}^{p} \frac{\partial^2 \rho}{\partial x_i^2} \\
&= \tilde{G}''(\rho) + \frac{p-1}{\rho} \tilde{G}'(\rho) \,.
\end{aligned}
$$

We can easily solve[8] this differential equation by separation of variables to find

$$
\tilde{G}(\rho) = a\rho + b, \quad \text{if } p = 1 \,,
$$

and

$$
\tilde{G}'(\rho) = \frac{\text{const.}}{\rho^{p-1}}, \quad \text{if } p \geq 2 \,.
$$

This, in turn, implies

$$
\tilde{G}(\rho) = a \ln \rho, \quad \text{if } p = 2 \,,
$$

and

$$
\tilde{G}(\rho) = \frac{a}{\rho^{p-2}}, \quad \text{if } p \geq 3 \,.
$$

We will focus on the last case but the reader should explore the cases $p = 1, 2$ on his or her own to get a better understanding. To find the undetermined constant, we integrate the defining equation over a ball of radius ϵ centered at the point x_0:

$$
\int_{B_\epsilon^p(x_0)} \Delta_p \tilde{G} \, d^p x = \int_{B_\epsilon^p(x_0)} \delta(x - x_0) \, d^p x \,.
$$

---

[8]See also the result (2.19) of Chapter 2.

By the properties of the Dirac delta function, the right-hand side is
1. The left-hand side can be rewritten by the use of the divergence
theorem:

$$\int_{B_\epsilon^p(x_0)} \Delta_p \tilde{G}\, d^p x = \int_{S_\epsilon^{p-1}(x_0)} \nabla_p \tilde{G} \cdot \xi \left(\epsilon^{p-1}\, d\Omega_{p-1} \right)$$

$$= \epsilon^{p-1} \int_{S_\epsilon^{p-1}(x_0)} \frac{d\tilde{G}}{d\rho}\, d\Omega_{p-1}$$

$$= \epsilon^{p-1} \int_{S_\epsilon^{p-1}(x_0)} \frac{a\,(2-p)}{\epsilon^{p-1}}\, d\Omega_{p-1}$$

$$= a\,(2-p)\,\Omega_{p-1}\,.$$

Hence

$$a = \frac{1}{(2-p)\,\Omega_{p-1}}\,.$$

Of course, differential equations come with boundary conditions.
So, let's modify the previous problem as follows. Let's seek the
Green's function which satisfies the same equation

$$\Delta_p G = \delta(x - x_0)\,, \tag{4.50}$$

for all $x \in B_p(0)$ and subjected to the boundary condition

$$G(\xi) = 0,\ \text{for all}\ \xi \in S^{p-1}\,. \tag{4.51}$$

Equation (4.50) now describes the electric potential caused by a point
charge at x_0 and an ideal conducting sphere with center at the origin.

To construct G, we write

$$G(x; x_0) = \tilde{G}(\rho) + g = \frac{1}{(2-p)\Omega_{p-1}\rho^{p-2}} + g\,,$$

and require that g is harmonic in B_p

$$\Delta_p g = 0$$

and cancel $\tilde{G}$ on the boundary of B_p:

$$g(\xi) = -\tilde{G}(\xi) = \frac{1}{(p-2)\Omega_{p-1}\rho^{p-2}} \text{ for all } \xi \in S^{p-1}.$$

In fact, the functional expression of g is identical to that of $\tilde{G}$. However, there are two parameters that have to be fixed: the location of the singular point representing a point charge and the strength of the charge. The location of the singular point of g cannot be inside $B_p(0)$. We will place the charge at the symmetric point x_0' to x_0 with respect to the sphere, i.e., the point x_0' that lies on the line passing through the origin and x_0 with $|x_0'| = 1/|x_0|$. We see then that

$$g(\rho') \propto \frac{1}{(2-p)\Omega_{p-1}\rho'^{p-2}},$$

where $\rho' = |x - x_0'|$. We must choose the charge to be of the correct strength so that G vanishes on the sphere. We easily see that the correct choice of g is

$$g(\rho') = -\frac{1}{(2-p)\Omega_{p-1}(|x_0|\rho')^{p-2}},$$

so that

$$G(x; x_0) = \frac{1}{(2-p)\Omega_{p-1}} \left(\frac{1}{\rho^{p-2}} - \frac{1}{(|x_0|\rho')^{p-2}} \right). \qquad (4.52)$$

Indeed, (4.52) satisfies Laplace's equation. Using the law of cosines and letting θ be the angle between the ray from the origin to x and the ray from the origin to x_0,

$$\rho = \sqrt{|x|^2 + |x_0|^2 - 2|x|\,|x_0|\cos\theta},$$

$$\rho' = \sqrt{|x|^2 + |x_0'|^2 - 2|x|\,|x_0'|\cos\theta} = \sqrt{|x|^2 + \frac{1}{|x_0|^2} - 2\frac{|x|}{|x_0|}\cos\theta},$$

we see that (4.52) vanishes on the unit sphere, i.e., when $|x| = 1$. The method of constructing G as described above is known as *the method of images*. Physicists use it routinely without paying attention to the mathematical details! The reason is that if a solution is found for a boundary problem, it must be unique.

Let's now return to the problem stated on page 108 and present an alternative solution using the results on the Green's function for the Laplace equation.

Alternative Solution Green's theorem (as shown in the footnote of page 73) for the function G and V,

$$\int_{B^p} (V\Delta_p G - G\Delta_p V)\, d^p x = \int_{S^{p-1}} \left(V\frac{\partial G}{\partial \xi} - G\frac{\partial V}{\partial \xi} \right) d\Omega_{p-1},$$

reduces to

$$\int_{B^p} V\delta(x - x_0)\, d^p x = \int_{S^{p-1}} V\frac{\partial G}{\partial \xi}\, d\Omega_{p-1},$$

since V is harmonic and G obeys (4.50) and (4.51). Since the left-hand side of the above equation becomes

$$\int_{B^p} V(x)\, \delta(x - x_0)\, d^p x = V(x_0),$$

and since

$$\frac{\partial G}{\partial \xi} = \frac{\partial G}{\partial |x|}\bigg|_{|x|=1} = \frac{1 - |x_0|^2}{\Omega_{p-1}(1 + |x_0|^2 - 2|x_0|\cos\theta)^{p/2}},$$

we can write the solution,

$$V(x_0) = \frac{1}{\Omega_{p-1}} \int_{\xi \in S^{p-1}} f(\xi)\, \frac{1 - |x_0|^2}{(1 + |x_0|^2 - 2|x_0|\cos\theta)^{p/2}}\, d\Omega_{p-1}. \quad (4.53)$$

Obviously, this gives the potential as an integral representation. ∎

Now that we have solved the problem using two alternative methods, knowing that the solution to the boundary-value problem is unique, we can equate the two solutions to arrive at the following result.

Theorem 4.28 *In $\mathbb{E}^p$ we have*

$$\sum_{n=0}^{\infty} r^n N(p,n) P_n(t) = \frac{1-r^2}{(1-2rt+r^2)^{p/2}}. \qquad (4.54)$$

Proof Let $x_0 = |x_0|\eta$. Starting from equation (4.47),

$$V(x_0) = \sum_{n,j} |x_0|^n Y_{n,j}(\eta) \int_{\xi \in S^{p-1}} f(\xi) Y_{n,j}(\xi)\, d\Omega_{p-1}$$

$$= \sum_{n=0}^{\infty} |x_0|^n \int_{\xi \in S^{p-1}} f(\xi) \sum_{j=1}^{N(p,n)} Y_{n,j}(\xi) Y_{n,j}(\eta)\, d\Omega_{p-1},$$

we rewrite it using Theorem 4.11,

$$V(x_0) = \sum_{n=0}^{\infty} |x_0|^n \int_{\xi \in S^{p-1}} f(\xi) \frac{N(p,n)}{\Omega_{p-1}} P_n(\langle \xi, \eta \rangle)\, d\Omega_{p-1}$$

$$= \frac{1}{\Omega_{p-1}} \int_{\xi \in S^{p-1}} f(\xi) \sum_{n=0}^{\infty} |x_0|^n N(p,n) P_n(\cos\theta)\, d\Omega_{p-1}.$$

Since the function f is arbitrary, we can compare the above equation with (4.53) and set $r = |x_0|$ and $t = \cos\theta$ to complete the proof. ∎

This concludes our development of spherical harmonics in p dimensions. We have briefly considered an application to boundary value problems; we will not delve further into applications. We have achieved our main goal to study the theory of spherical harmonics and the corresponding Legendre polynomials in $\mathbb{E}^p$. We urge the reader to seek out applications on his or her own. Perhaps search for instances in physics where spherical harmonics are used in $\mathbb{E}^3$ and try to generalize to p dimensions. One could start with the multipole expansion of an electrostatic field (see [10], [11]) or the wave function of an electron in a hydrogenic atom (see [17], [12]).

4.5   Problems

1. For a homogeneous function[9] $f(x_1, x_2, \ldots, x_p)$ of degree n prove that

$$\sum_{i_1, i_2, \ldots, i_\ell = 1}^{p} \frac{\partial^\ell f}{\partial x_{i_1} \partial x_{i_2} \ldots \partial x_{i_\ell}} x_{i_1} x_{i_2} \ldots x_{i_\ell} = \frac{n!}{(n-\ell)!} f .$$

2. Prove that the function

$$f(r\xi) = \frac{1 - r^2}{(1 - 2r \langle \xi, \eta \rangle + r^2)^{p/2}}$$

is harmonic.

3. Prove equation (4.39).

4. In $p = 3$, show that

$$\frac{1}{|\vec{r} - \vec{r}'|} = \frac{1}{r_>} \sum_{n=0}^{\infty} \left(\frac{r_<}{r_>} \right)^n P_n(\cos\gamma) ,$$

where γ is the angle between the two vectors $\vec{r}, \vec{r}'$ and $r_> = \max\{r, r'\}$ while $r_< = \min\{r, r'\}$.

As a consequence, show that, for $p = 3$, equation (4.54) simplifies to

$$\sum_{n=0}^{\infty} r^n P_n(t) = \frac{1}{(1 - 2rt + r^2)^{1/2}} .$$

5. Generalize the previous problem to p dimensions. That is, given the vectors x and y in $\mathbb{E}^n$, expand $1/|x - y|$ in terms of the Legendre polynomials.

---

[9]Like a homogeneous polynomial, a function is homogeneous of degree n provided $f(tx_1, tx_2, \ldots, tx_p) = t^n f(x_1, x_2, \ldots, x_p)$.

6. Use Problem 7 of Chapter 2 to find a relation between the spherical harmonics in p and $p-1$ dimensions. More precisely, set

$$Y_n = \Theta \tilde{Y}_{n'},$$

where Θ is a function of θ_{p-2} and $Y_n, \tilde{Y}_{n'}$ are spherical harmonics on the $(p-1)$-dimensional and $(p-2)$-dimensional spheres respectively. Then show that

$$\Theta(\theta) = (\sin\theta)^{-\frac{p-3}{2}} P_{n+\frac{p-3}{2}}^{n'-\frac{p-3}{2}}(\cos\theta),$$

where $P_\ell^m(t)$ are the associated Legendre polynomials. Using this result, express the spherical harmonics in p dimensions in terms of those in $p = 2$ dimensions.

Chapter 5

Solutions to Problems

5.1 Solutions to Problems of Chapter 2

1. If f is harmonic, then

$$\Delta_p f = \frac{\partial^2 f}{\partial r^2} + \frac{p-1}{r}\frac{\partial f}{\partial r} + \frac{1}{r^2}\Delta_{S^{p-1}} f = 0.$$

We must now show that $\Delta_p g = 0$.

$$
\begin{aligned}
\Delta_p g &= \left[\frac{\partial^2}{\partial r^2} + \frac{p-1}{r}\frac{\partial}{\partial r} + \frac{1}{r^2}\Delta_{S^{p-1}}\right] r\frac{\partial f}{\partial r} \\
&= \left(r\frac{\partial^3 f}{\partial r^3} + 2\frac{\partial^2 f}{\partial r^2}\right) + \frac{p-1}{r}\left(\frac{\partial f}{\partial r} + r\frac{\partial^2 f}{\partial r^2}\right) \\
&\quad + \frac{1}{r}\frac{\partial}{\partial r}\Delta_{S^{p-1}} f .
\end{aligned}
$$

The last term in the right-hand side can be computed from the first equation:

$$
\begin{aligned}
\frac{\partial}{\partial r}\Delta_{S^{p-1}} f &= -\frac{\partial}{\partial r}\left(r^2\frac{\partial^2 f}{\partial r^2} + (p-1)r\frac{\partial f}{\partial r}\right) \\
&= -r^2\frac{\partial^3 f}{\partial r^3} - 2r\frac{\partial^2 f}{\partial r^2} - (p-1)\frac{\partial f}{\partial r} - (p-1)r\frac{\partial^2 f}{\partial r^2}.
\end{aligned}
$$

Upon substituting in the expression for $\Delta_p g$ we find that all the terms add to zero — i.e. g is harmonic. ∎

2. We set $x_i = r\,\xi_i$. Then

$$V_p = r^p \int \cdots \int_{\xi_1^2 + \xi_2^2 + \cdots + \xi_p^2 \leq 1} d\xi_1\, d\xi_2 \cdots d\xi_p\,,$$

which is exactly the relation $V_p = r^p\, U_p$. We now write the integral U_p in the form

$$U_p = \int_{-1}^{1} d\xi_p \int \cdots \int_{\xi_1^2 + \xi_2^2 + \cdots + \xi_{p-1}^2 \leq 1 - \xi_p^2} d\xi_1\, d\xi_2 \cdots d\xi_{p-1}\,.$$

We notice that the multiple integral is over a $(p-1)$-ball of radius $(1 - \xi_p^2)^{1/2}$. So, similarly to the first substitution, we write

$$\xi_i = (1 - \xi_p^2)^{1/2}\, \eta_i\,, \quad i = 1, 2, \ldots, p-1\,.$$

Hence the right-hand side can be separated in a product of two integrals:

$$U_p = \int_{-1}^{1} d\xi_p\, (1 - \xi_p^2)^{\frac{p-1}{2}} \int \cdots \int_{\eta_1^2 + \eta_2^2 + \cdots + \eta_{p-1}^2 \leq 1} d\eta_1\, d\eta_2 \cdots d\eta_{p-1}\,.$$

Obviously, the multiple integral is equal to U_{p-1} resulting in the recurrence relation

$$U_p = U_{p-1} \int_{-1}^{1} dt\, (1 - t^2)^{\frac{p-1}{2}}\,.$$

We could write this integral as a beta function (through its definition) and consequently as a gamma function (through the property relating the two functions). However, this is not necessary[1] since we can easily use elementary methods. Using the

---

[1]The reader may want to solve the problem using the definition of the beta

substitution $t = \sin\alpha$,

$$U_p \;=\; U_{p-1} \int_{-1}^{1} d(\sin\alpha)\,(\cos\alpha)^{p-1} \;=\; U_{p-1} \underbrace{\int_{-\pi/2}^{\pi/2} (\cos\alpha)^p\, d\alpha}_{c_p}\;.$$

We have kept the first form of the integral since it suggests to evaluate it using integration by parts:

$$c_p \;=\; \underbrace{\sin\alpha\,(\cos\alpha)^{p-1}\Big|_{-\pi/2}^{\pi/2}}_{\text{vanishes}} + (p-1)\int_{-\pi/2}^{\pi/2} \sin^2\alpha\,(\cos\alpha)^{p-2}\, d\alpha$$

$$\;=\; (p-1)\int_{-\pi/2}^{\pi/2} (1-\cos^2\alpha)\,(\cos\alpha)^{p-2}\, d\alpha$$

$$\;=\; (p-1)\int_{-\pi/2}^{\pi/2} (\cos\alpha)^{p-2}\, d\alpha - (p-1)\int_{-\pi/2}^{\pi/2} (\cos\alpha)^{p}\, d\alpha$$

$$\;=\; (p-1)\,c_{p-2} - (p-1)\,c_p\,,$$

or

$$c_p \;=\; \frac{p-1}{p}\,c_{p-2}\,.$$

Inductively, if p is even,

$$c_p \;=\; \frac{p-1}{p}\frac{p-3}{p-2}\frac{p-5}{p-4}\cdots\frac{1}{2}\,c_0 \;=\; \frac{(p-1)!!}{p!!}\,\pi\,,$$

function

$$B(x,y) \;=\; \int_0^1 t^{x-1}(1-t)^{y-1}\, dt\,,$$

and the property

$$B(x,y) \;=\; \frac{\Gamma(x)\,\Gamma(y)}{\Gamma(x+y)}\,.$$

and if p is odd,

$$c_p = \frac{p-1}{p}\frac{p-3}{p-2}\frac{p-5}{p-4}\cdots\frac{2}{3}c_1 = \frac{(p-1)!!}{p!!}2.$$

Using the relation $U_p = c_p U_{p-1}$ inductively, we finally find

$$U_p = c_p\,c_{p-1}\,c_{p-2}\cdots c_1\,U_0.$$

Let's assume that p is even (the reader should do the case in which p is odd). Then

$$U_p = \left[\frac{(p-1)!!}{p!!}\pi\right]\left[\frac{(p-2)!!}{(p-1)!!}2\right]\left[\frac{(p-3)!!}{(p-2)!!}\pi\right]\cdots\left[\frac{1!!}{2!!}\pi\right]\left[\frac{0!!}{1!!}2\right]2$$

$$= \frac{2^{\frac{p}{2}}\pi^{\frac{p}{2}}}{p!!}.$$

From the property $\Gamma(x+1) = x\,\Gamma(x)$, which the reader can prove from (2.5) using integration by parts, it follows that for any natural number n, $\Gamma(n+1) = n!$. Thus, for even p,

$$\Gamma\left(\frac{p}{2}+1\right) = \left(\frac{p}{2}\right)! = \frac{p!!}{2^{\frac{p}{2}}},$$

which implies

$$U_p = \frac{\pi^{\frac{p}{2}}}{\Gamma\left(\frac{p}{2}+1\right)}.$$

From the relation

$$A_{p-1} = \frac{dV_p}{dr} = \frac{p\,\pi^{\frac{p}{2}}}{\Gamma\left(\frac{p}{2}+1\right)}r^{p-1},$$

we conclude that

$$\Omega_{p-1} = \frac{p\,\pi^{\frac{p}{2}}}{\Gamma\left(\frac{p}{2}+1\right)} = \frac{p\,\pi^{\frac{p}{2}}}{\frac{p}{2}\Gamma\left(\frac{p}{2}\right)} = \frac{2\pi^{\frac{p}{2}}}{\Gamma\left(\frac{p}{2}\right)},$$

as found previously in Chapter 2.   ■

3. Comparing the result (2.10) with the defining expression of the metric tensor (2.20), we see that for a sphere the metric tensor is diagonal, i.e. $g_{ij} = 0$ for $i \neq j$. It is also easy to see that

$$
\begin{aligned}
g_{00} &= \sin^2 \theta_{p-2} \sin^2 \theta_{p-3} \cdots \sin^2 \theta_2 \sin^2 \theta_1, \\
g_{11} &= \sin^2 \theta_{p-2} \sin^2 \theta_{p-3} \cdots \sin^2 \theta_2,
\end{aligned}
$$

$$
\vdots
$$

$$
\begin{aligned}
g_{p-4,p-4} &= \sin^2 \theta_{p-2} \sin^2 \theta_{p-3}, \\
g_{p-3,p-3} &= \sin^2 \theta_{p-2}, \\
g_{p-2,p-2} &= 1.
\end{aligned}
$$

where we used the index 0 for the ϕ-coordinate. Since the matrix $[g_{ij}]$ is diagonal

$$
\begin{aligned}
g &= \det[g_{ij}] = \prod_{i=0}^{p-2} g_{ii} \\
&= (\sin \theta_{p-2})^{2(p-2)} (\sin \theta_{p-3})^{2(p-3)} \cdots (\sin \theta_2)^{2 \cdot 2} (\sin \theta_1)^{2 \cdot 1}.
\end{aligned}
$$

The vanishing diagonal elements of the metric tensor also imply that the spherical coordinates are orthogonal. Therefore an infinitesimal box of length $\sqrt{g_{ii}}\, d\theta_i$, $i = 0, 1, \ldots, p-2$, along the i-th direction has an infinitesimal volume

$$
d\Omega_{p-1} = \prod_{i=0}^{p-2} \sqrt{g_{ii}}\, d\theta_i = \sqrt{g}\, d\theta_{p-2} \cdots d\theta_1 d\phi.
$$

The total volume of the sphere is found by adding all infinitesimal boxes that make the sphere, that is, by integrating over the range of the coordinates. This gives the integral asked to prove and which splits to $p-1$ independent integrals:

$$
\Omega_{p-1} = 2\pi \prod_{i=1}^{p-2} 2 \int_0^{\pi/2} (\sin \theta_i)^i\, d\theta_i = 2\pi \prod_{i=1}^{p-2} \int_{-\pi/2}^{\pi/2} (\cos \alpha)^i\, d\alpha.
$$

The '2' in front of the integrals is due to the fact that the actual range is from $-\pi/2$ to $\pi/2$ but the infinitesimal volumes are positive even for the negative range. (More precisely, $\sqrt{\sin^2\theta} = |\sin\theta|$.) Therefore,

$$\Omega_{p-1} = 2\pi\, c_{p-2}\, c_{p-3}\cdots c_2\, c_1\,,$$

where the numbers c_i have been computed in the previous problem. The reader should finish the problem by making sure that the above product results in the expected expression. ∎

4. The volume bound by the $(p-1)$-ellipsoid is

$$V_p = \int\cdots\int_{\frac{x_1^2}{a_1^2}+\frac{x_2^2}{a_2^2}+\cdots+\frac{x_p^2}{a_p^2}\leq 1} dx_1\, dx_2\cdots dx_p\,.$$

If we make the substitution of variables $y_i = x_i/a_i$, $i = 1, 2, ..., p$, the integral becomes that for the volume of a unit p-ball,

$$V_p = a_1 a_2\cdots a_p \int\cdots\int_{y_1^2+y_2^2+\cdots+y_p^2\leq 1} dy_1\, dy_2\cdots dy_p\,,$$

and thus

$$V_p = \frac{\pi^{\frac{p}{2}}}{\Gamma\left(\frac{p}{2}+1\right)}\, a_1 a_2\cdots a_p\,.$$

The reader should examine whether the same argument works for the surface area of the ellipsoid. S/he can try to experiment with the circle and the ellipse as they are simpler than the general case. ∎

5. Stirling's approximation formula for the gamma function is

$$\Gamma(x+1) = \sqrt{2\pi x}\, x^x\, e^{-x}\left(1 + \frac{1}{12}\frac{1}{x} + \cdots\right).$$

For very large values of x,

$$\Gamma(x+1) \simeq \sqrt{2\pi x}\, x^x\, e^{-x}\,,$$

with the error getting smaller as x increases. Let $x = p/2$. Then

$$\lim_{p\to\infty} U_p = \lim_{x\to\infty} \frac{1}{\sqrt{2\pi x}} \cdot \lim_{x\to\infty} \left(\frac{e\pi}{x}\right)^x = 0.$$

Similarly, $\Omega_{p-1} \to 0$.

In Problem 2, we showed that $U_p = c_p U_{p-1}$ and we found a formula for the coefficients c_p. In particular,

$$c_1 = 2, \quad c_2 = \frac{\pi}{2}, \quad c_3 = \frac{4}{3}, \quad c_4 = \frac{3\pi}{8},$$

$$c_5 = \frac{16}{15}, \quad c_6 = \frac{5\pi}{16}, \quad c_7 = \frac{32}{35}, \quad \cdots$$

Notice that the sequence up to c_5 has values greater than 1. Both c_6 and c_7 are less than one. Any terms following them get progressively smaller:

$$c_p = \left(1 - \frac{1}{p}\right) c_{p-2} \le c_{p-2} < 1, \quad p \ge 6.$$

The corresponding sequence of volumes is:

$$U_0 = 1, \quad U_1 = 2, \quad U_2 = \pi, \quad U_3 = \frac{4\pi}{3},$$

$$U_4 = \frac{\pi^2}{2}, \quad U_5 = \frac{8\pi^2}{15}, \quad U_6 = \frac{\pi^3}{6}, \quad \cdots$$

The volumes increase up to U_5; then they start to decrease. Therefore, the volume U_5 is the greatest.

Since $A_p = dV_{p+1}/dr = (p+1)\, r^p\, U_{p+1}$, we have $\Omega_p = (p+1)\, U_{p+1}$ and thus

$$\frac{\Omega_p}{\Omega_{p-1}} = \frac{p+1}{p} \frac{U_{p+1}}{U_p},$$

or

$$\Omega_p = \frac{p+1}{p} c_{p+1} \Omega_{p-1} = c_{p-1} \Omega_{p-1}.$$

The sequence of the areas is then:

$$\Omega_0 = 2, \quad \Omega_1 = 2\pi, \quad \Omega_2 = 4\pi, \quad \Omega_3 = 2\pi^2, \quad \Omega_4 = \frac{8\pi^2}{3},$$

$$\Omega_5 = \pi^3, \quad \Omega_6 = \frac{16\pi^3}{15}, \quad \Omega_7 = \frac{\pi^4}{3}, \quad \Omega_8 = \frac{32\pi^4}{105}, \quad \dots$$

It is now seen that Ω_6 is the largest term of the sequence. ∎

6. We have found the formula for the volume of the p-ball. However, to do the calculation in this problem we need the volume of the p-cube too. Without loss of generality, since we will consider ratios, we will normalize the objects such that we use unit balls in both cases.

Given a unit ball, the edge of the circumscribed cube is 2 and its volume $V_c = 2^p$. Therefore the ratio R_c of the volume of the p-ball over the volume of the circumscribed p-cube is

$$R_c = \frac{V}{V_c} = \frac{\pi^{p/2}}{2^p \, \Gamma\left(\frac{p}{2} + 1\right)}.$$

This ratio measures how well the p-ball fits in the p-cube: The closer the ratio to 1, the better the fit.

Given a unit ball, the diagonal of the inscribed cube is 2. If a is its edge, by the Pythagorean theorem, $2^2 = p\,a^2$ and thus $a = 2/\sqrt{p}$ resulting in a volume $V_i = 2^p/p^{p/2}$. Therefore the ratio R_i of the volume of the inscribed p-cube over the volume of the p-ball is

$$R_i = \frac{V_i}{V} = \frac{2^p \, \Gamma\left(\frac{p}{2} + 1\right)}{p^{p/2} \, \pi^{p/2}}.$$

This ratio measures how well the p-cube fits in the p-ball: The closer the ratio to 1, the better the fit.

Singmaster's theorem claims that $R_c(p) \geq R_i(p)$ if and only if $p \leq 8$. We will show it using the same approach we used to show that the sphere areas and the ball volume exhibit a maximum. That is, we will first show that the ratio R_c/R_i approaches zero as $p \to \infty$. Then by computing the first terms of the sequence, we will show that there is a change of behavior at $p = 8$.

By Stirling's approximation, we can find

$$\frac{R_c(p)}{R_i(p)} = \frac{p^{p/2}\,\pi^p}{2^{2p}\,\Gamma\left(\frac{p}{2}+1\right)^2} \simeq \frac{e^2}{\pi(p+2)}\left[\frac{\pi\,e\,\sqrt{p}}{2\,(p+2)}\right]^p$$

$$\leq \frac{e^2}{\pi p}\left[\frac{\pi\,e}{2}\,\frac{1}{\sqrt{p}}\right]^p.$$

The right-hand side converges to 0 and therefore $R_c/R_i \to 0$ as $p \to \infty$. The reader can fill the remaining details. Although not challenging, they require some labor. ∎

7. Once more, look at the definition of spherical coordinates. Notice that the equations that define S^{p-2} as a subspace of S^{p-1} are similar to the equations that define S^{p-1} as a subspace of R^p. Here the θ_{p-2} is considered radial (it is orthogonal to all other directions). Also, notice that S^{p-2} is not a unit sphere but has radius $\sin\theta_{p-2}$. Then we use the decomposition (2.17) to find the equation sought between $\Delta_{S^{p-1}}$ and $\Delta_{S^{p-2}}$. If you find this argument a little imprecise for your taste, try to redo the computation of Section 2.6 or, perhaps, use the general expression (2.15). Once the recursion relation is found, the explicit expression of $\Delta_{S^{p-1}}$ in terms of the angular coordinates is found easily by induction. ∎

5.2   Solutions to Problems of Chapter 3

1. Using the definition of $B_{k,n}(x)$, we have

$$x\,B_{k-1,n-1}(x) + (1-x)\,B_{k,n-1}(x) =$$
$$\left[\binom{n-1}{k-1} + \binom{n-1}{k}\right] x^k\,(1-x)^{n-k}.$$

Using the definition (3.8) it follows that

$$\binom{n-1}{k-1} + \binom{n-1}{k} = \binom{n}{k},$$

and hence

$$x\,B_{k-1,n-1}(x) + (1-x)\,B_{k,n-1}(x) =$$
$$\binom{n}{k}\,x^k\,(1-x)^{n-k} = B_{k,n}(x)\,.$$

Now we notice that

$$\frac{n-k}{n}\binom{n}{k} = \frac{n-k}{n}\frac{n!}{k!(n-k)!}$$
$$= \frac{(n-1)!}{k!(n-1-k)!} = \binom{n-1}{k}\,.$$

Similarly

$$\frac{k+1}{n}\binom{n}{k+1} = \binom{n-1}{k}\,.$$

Therefore the sum

$$\frac{n-k}{n}\,B_{k,n}(x) + \frac{k+1}{n}\,B_{k+1,n}(x)$$

is equal to

$$\binom{n-1}{k}\left[x^k\,(1-x)^{n-k} + x^{k+1}\,(1-x)^{n-k-1}\right] = B_{k,n-1}(x)\,.$$

Differentiating $B_{k,n}(x)$ we find

$$B'_{k,n}(x) = \binom{n}{k}\,k\,x^{k-1}\,(1-x)^{n-k} - \binom{n}{k}\,x^k\,(n-k)\,(1-x)^{n-k-1}$$
$$= n\binom{n-1}{k-1}x^{k-1}\,(1-x)^{n-k} - n\binom{n-1}{k}x^k\,(1-x)^{n-k-1}$$
$$= n\,(B_{k-1,n-1}(x) - B_{k,n-1}(x))\,,$$

as sought.   ∎

2. Differentiating equation (3.15) with respect to x and multiplying by x/n, we find

$$\frac{x}{n}(x+y)^{n-3}\left[n x^2 + xy\left(3-\frac{1}{n}\right)+\frac{y^2}{n}\right] =$$

$$\sum_{k=0}^{n}\binom{n}{k}\left(\frac{k}{n}\right)^3 x^k y^{n-k}.$$

Finally, substituting $y = 1 - x$ gives

$$B_n(x;x^3) = \left(1-\frac{3}{n}+\frac{2}{n^2}\right)x^3 + \left(\frac{3}{n}-\frac{3}{n^2}\right)x^2+\frac{x}{n^2},$$

which are the polynomials we wished to find.

Incidentally, notice that we can continue the same way to construct the polynomials $B_n(x;x^m)$ for any natural number m.
∎

3. Computing $\frac{dB_n(x;f)}{dx}$, we find

$$\sum_{k=0}^{n}\binom{n}{k}f\left(\frac{k}{n}\right)\left[k\,x^{k-1}(1-x)^{n-k}-(n-k)\,x^k(1-x)^{n-k-1}\right]$$

or equivalently

$$\sum_{k=1}^{n}\binom{n}{k}f\left(\frac{k}{n}\right)k\,x^{k-1}(1-x)^{n-k}$$

$$-\sum_{k=0}^{n-1}\binom{n}{k}f\left(\frac{k}{n}\right)(n-k)\,x^k(1-x)^{n-k-1}.$$

Replacing k by $j+1$ in the first sum, it becomes

$$\sum_{j=0}^{n-1}\frac{n!}{j!\,(n-j-1)!}f\left(\frac{j+1}{n}\right)(j+1)\,x^j(1-x)^{n-j-1}$$

or

$$n \sum_{k=0}^{n-1} \frac{(n-1)!}{k!(n-k-1)!} f\left(\frac{k+1}{n}\right) x^k (1-x)^{n-k-1}.$$

Meanwhile the second sum becomes

$$-\sum_{k=0}^{n-1} \frac{n!}{k!(n-k-1)!} f\left(\frac{k}{n}\right) x^k (1-x)^{n-k-1}$$

or

$$-n \sum_{k=0}^{n-1} \binom{n-1}{k} f\left(\frac{k}{n}\right) x^k (1-x)^{n-k-1}.$$

Putting them back together we find

$$\frac{dB_n(x;f)}{dx} = \sum_{k=0}^{n-1} \left[\frac{f\left(\frac{k+1}{n}\right) - f\left(\frac{k}{n}\right)}{\left(\frac{k+1}{n}\right) - \left(\frac{k}{n}\right)}\right] \binom{n-1}{k} x^k (1-x)^{n-1-k}.$$

Thus, by the same argument used to derive (3.18), the error

$$E = \left| f'(x) - \frac{dB_n(x;f)}{dx} \right|$$

satisfies the inequality

$$E \le \sum_{k=0}^{n-1} \left| f'(x) - \frac{f\left(\frac{k+1}{n}\right) - f\left(\frac{k}{n}\right)}{\left(\frac{k+1}{n}\right) - \left(\frac{k}{n}\right)} \right| \binom{n-1}{k} x^k (1-x)^{n-1-k}.$$

Now, let $\epsilon > 0$ be arbitrary. Since $f'(x)$ is, by assumption, continuous on the closed and bounded interval $[0,1]$, it follows that $f'(x)$ is uniformly continuous on $[0,1]$; consequently, $f(x)$ is uniformly differentiable on $[0,1]$, so we can choose n large enough to ensure that for any k,

$$\left| f'\left(\frac{k}{n-1}\right) - \frac{f\left(\frac{k+1}{n}\right) - f\left(\frac{k}{n}\right)}{\left(\frac{k+1}{n}\right) - \left(\frac{k}{n}\right)} \right| < \frac{\epsilon}{2}.$$

Thus,

$$E < \frac{\epsilon}{2} + \sum_{k=0}^{n-1} \left| f'(x) - f'\left(\frac{k}{n-1}\right) \right| \binom{n-1}{k} x^k (1-x)^{n-1-k}.$$

But $f'(x)$ is continuous, so the remaining sum is analogous to that of (3.18) with f replaced by f' and n replaced by $n-1$. Thus, by the same argument which led to (3.19), it follows that for large enough n we can guarantee that for all $x \in [0,1]$,

$$\sum_{k=0}^{n-1} \left| f'(x) - f'\left(\frac{k}{n-1}\right) \right| \binom{n-1}{k} x^k (1-x)^{n-1-k} < \frac{\epsilon}{2},$$

leaving us with $E < \epsilon$. This proves that $\frac{dB_n(x;f)}{dx}$ converges to $f'(x)$ uniformly on $[0,1]$ as $n \to \infty$. ∎

5.3 Solutions to Problems of Chapter 4

1. The proof is an extension of that used in the derivation of Euler's equation (4.1). In the definition of the homogeneous function,

$$f_n(tx_1, tx_2, \ldots, tx_p) = t^n f_n(x_1, x_2, \ldots, x_p),$$

let's set $u_i = tx_i$, for all i and differentiate the defining equation with respect to t:

$$\sum_{i=1}^{p} \frac{\partial f_n(u_1, u_2, \ldots, u_p)}{\partial u_i} x_i = n t^{n-1} f_n(x_1, x_2, \ldots, x_p).$$

Differentiating a second time

$$\sum_{j,i=1}^{p} \frac{\partial^2 f_n(u_1, u_2, \ldots, u_p)}{\partial u_j \partial u_i} x_j x_i = n(n-1) t^{n-2} f_n(x_1, x_2, \ldots, x_p).$$

If we differentiate $\ell - 2$ times more and then set $t = 1$ we find the advertised equation. ∎

2. This statement is immediate from equation (4.54). The left-hand side of this equation is

$$\sum_{n=0}^{\infty} r^n \, N(p,n) \, P_n(t) \;=\; \sum_{n=0}^{\infty} r^n \, N(p,n) \, L_n(\xi)$$

$$=\; \sum_{n=0}^{\infty} N(p,n) \, L_n(r\xi) \,,$$

that is, a sum of harmonic functions. Hence the sum and thus the right-hand side must be a harmonic function too.

The reader is invited to show this result by a direct calculation. In particular, using the spherical symmetry, the vector η can be oriented along the x_p-axis. Then

$$f(r\xi) \;=\; \frac{1 - r^2}{(1 - 2r\,\cos\theta + r^2)^{p/2}} \,,$$

where for simplicity we have set $\theta = \theta_{p-2}$. We can then use expressions (2.17) and (2.21),

$$\Delta_p \;=\; \frac{\partial^2}{\partial r^2} + \frac{p-1}{r}\frac{\partial}{\partial r} + \frac{1}{r^2(\sin\theta)^{p-2}}\frac{\partial}{\partial\theta}\left[(\sin\theta)^{p-2}\frac{\partial}{\partial\theta}\right]$$
$$+ \frac{1}{r^2(\sin\theta)^2}\,\Delta_{S^{p-2}} \,,$$

to compute $\Delta_p f$ explicitly. ∎

3. Let

$$C_n^{\left(\frac{p-2}{2}\right)}(t) \;=\; a_n \, P_n(t) \,,$$

where a_n may possibly depend on p too. For $n = 0$, the corresponding polynomials coincide; hence $a_0 = 1$. Substituting in (4.38)

$$(n+1)\,a_{n+1}\,P_{n+1}(t) - (2n + p - 2)\,a_n\,t\,P_n(t)$$
$$+ (n + p - 3)\,a_{n-1}\,P_{n-1}(t) \;=\; 0 \,.$$

This equation will be identical to equation (4.36) if

$$
\begin{aligned}
(n+1)\, a_{n+1} &= b_n\,(n+p-2),\\
(2n+p-2)\, a_n &= b_n\,(2n+p-2),\\
(n+p-3)\, a_{n-1} &= b_n\, n,
\end{aligned}
$$

where b_n is a possible nonzero common factor. It is immediate that $b_n = a_n$ and

$$
a_n = \frac{p+n-3}{n}\, a_{n-1},
$$

for all $n = 1, 2, \ldots.$ Recursively,

$$
\begin{aligned}
a_n &= \frac{p+n-3}{n}\,\frac{p+n-4}{n-1}\,\frac{p+n-5}{n-2}\cdots\frac{p-1}{2}\,\frac{p-2}{1}\, a_0\\
&= \frac{(n+p-3)!}{n!\,(p-3)!} = \binom{n+p-3}{n},
\end{aligned}
$$

which is the advertised result.   ■

4. Since the expression $|\vec{r} - \vec{r}\,'|$ is symmetric in the two vectors, without loss of generality, we will assume that $r' > r$. We then choose the orientation of the coordinate system such that the vector $\vec{r}\,'$ is along the x_3-axis. In this case the angle γ between the two vectors coincides with the θ-coordinate. As has been seen, the function

$$
f(r,\theta) = \frac{1}{|\vec{r} - \vec{r}\,'|} = \frac{1}{\sqrt{r^2 + r'^2 - 2rr'\cos\theta}},
$$

is harmonic and therefore it can be expanded in the form[2]

$$
\frac{1}{|\vec{r} - \vec{r}\,'|} = \sum_{n=0}^{\infty}\sum_{j=-n}^{n} c_{n,j}\, r^n\, Y_{n,j}(\phi,\theta).
$$

---

[2] The condition $r < r'$ requires that f is regular at $r = 0$. When $r > r'$ is assumed instead, the expansion should be in terms of r' and f should be regular at $r' = 0$.

The left-hand side of this equation does not depend on ϕ. Therefore[3] $c_{n,j} = 0$ for all $j \neq 0$ and the expansion simplifies to

$$\frac{1}{|\vec{r} - \vec{r}\,'|} = \sum_{n=0}^{\infty} c_{n,0}\, r^n\, Y_{n,0}(\phi, \theta) = \sum_{n=0}^{\infty} e_n\, r^n\, P_n(\cos\theta)\,,$$

where we have renamed the necessary coefficients as e_n. We can calculate these constants by considering the value of the function on the axis of symmetry, $\theta = 0$:

$$f(r, 0) = \sum_{n=0}^{\infty} e_n\, r^n\,.$$

At the same time

$$f(r, 0) = \frac{1}{r' - r} = \frac{1}{r'}\frac{1}{1 - (r/r')} = \frac{1}{r'} \sum_{n=0}^{\infty} \left(\frac{r}{r'}\right)^n\,.$$

Comparing the last two expressions, we conclude that $e_n = (r')^{-n-1}$ and thus

$$\frac{1}{|\vec{r} - \vec{r}\,'|} = \frac{1}{r'} \sum_{n=0}^{\infty} \left(\frac{r}{r'}\right)^n P_n(\cos\theta)\,,$$

as required.

The relation follows directly from the above

$$\frac{1}{\sqrt{1 + r^2 - 2r\cos\gamma}} = \sum_{n=0}^{\infty} r^n\, P_n(\cos\gamma)\,.$$

upon taking $r_< = r$ and $r_> = 1$. ∎

---

[3]To be concrete, we choose here to use the spherical harmonics in (4.21a)-(4.21b). In the next solution, we prove a more general result without making this choice.

5. We duplicate the argument of the previous problem. We thus assume that $r' > r$ and choose the orientation of the coordinate system such that the vector x' is along the x_p-axis. In this case the angle γ between the two vectors coincides with the θ_{p-2}-coordinate. As has been seen, the function

$$f(x) = \frac{1}{|x - x'|^{p-2}} = \frac{1}{(r^2 + r'^2 - 2rr' \cos\gamma)^{\frac{p-2}{2}}},$$

is harmonic and therefore it can be expanded in the form

$$\frac{1}{|x - x'|^{p-2}} = \sum_{n=0}^{\infty} \sum_{j} c_{n,j}\, r^n\, Y_{n,j}(\xi),$$

where $x = r\xi$. The left-hand side of this equation depends only on θ_{p-2}. Therefore $c_{n,j} = 0$ for all those spherical harmonics that contain dependence[4] on $\theta_{p-3}, \ldots, \theta_1, \phi$; as a result only the spherical harmonics that are proportional to $P_n(t)$ will survive:

$$\frac{1}{|x - x'|^{p-2}} = \sum_{n=0}^{\infty} e_n\, r^n\, P_n(\cos\gamma),$$

where we have renamed the necessary coefficients as e_n. We can calculate these constants by considering the value of the function on the axis of symmetry, $\gamma = 0$:

$$f(r, \gamma = 0) = \sum_{n=0}^{\infty} e_n\, r^n.$$

At the same time

$$f(r, \gamma = 0) = \frac{1}{(r' - r)^{p-2}} = \frac{1}{(r')^{p-2}}\left(1 - \frac{r}{r'}\right)^{2-p}$$

$$= \frac{1}{(r')^{p-2}} \sum_{n=0}^{\infty} (-1)^n \frac{(2-p)_n}{n!} \left(\frac{r}{r'}\right)^n,$$

---

[4]See Problem 6 of this chapter.

using the Taylor expansion of the binomial $(1 + x)^{\alpha}$. It can be checked that

$$(2 - p)_n = (-1)^n \frac{(p + n - 3)!}{(p - 3)!}.$$

Therefore

$$f(r, \gamma = 0) = \frac{1}{(r')^{p-2}} \sum_{n=0}^{\infty} \binom{n + p - 3}{n} \left(\frac{r}{r'}\right)^n.$$

Comparing the two expressions at $\gamma = 0$, we conclude that

$$e_n = \frac{1}{(r')^{p-2+n}} \binom{n + p - 3}{n},$$

and thus

$$\frac{1}{|x - x'|^{p-2}} = \frac{1}{(r')^{p-2}} \sum_{n=0}^{\infty} \binom{n + p - 3}{n} \left(\frac{r}{r'}\right)^n P_n(\cos \gamma),$$

the generalization of the 3-dimensional formula. Notice that, in the right-hand side, the Gegenbauer polynomials have appeared. If we set $r' = 1$, $r < 1$, and $t = \cos \gamma$, then

$$\frac{1}{(1 + r^2 - 2tr)^{\frac{p-2}{2}}} = \sum_{n=0}^{\infty} r^n C_n^{\left(\frac{p-2}{2}\right)}(t);$$

that is, we have found the generating function of the Gegenbauer polynomials. In fact, more generally,

$$\frac{1}{(1 + r^2 - 2tr)^{\alpha}} = \sum_{n=0}^{\infty} r^n C_n^{(\alpha)}(t).$$

The reader may want to prove it.   ∎

6. Since

$$\Delta_{S^{p-1}} Y_n = n(2 - p - n) Y_n,$$
$$\Delta_{S^{p-2}} \tilde{Y}_{n'} = n'(3 - p - n') \tilde{Y}_{n'},$$

the relation

$$\Delta_{S^{p-1}} Y_n = \frac{\tilde{Y}_{n'}}{(\sin\theta_{p-2})^{p-2}} \frac{d}{d\theta_{p-2}} \left[(\sin\theta_{p-2})^{p-2} \frac{d\Theta}{d\theta_{p-2}} \right]$$

$$+ \frac{\Theta}{(\sin\theta_{p-2})^2} \Delta_{S^{p-2}} \tilde{Y}_{n'} \,,$$

takes the form

$$\frac{d^2\Theta}{d\theta^2} + (p-2)\cot\theta \frac{d\Theta}{d\theta} + \left(\frac{A}{\sin^2\theta} - B \right) \Theta = 0 \,,$$

where

$$A = n'(3 - p - n'), \quad B = n(2 - p - n) \,,$$

and, for simplicity, we wrote θ instead of θ_{p-2}. A reader with experience in special functions will recognize immediately that the previous differential equation is *almost* the one for the associated Legendre polynomials — "almost" since the numerical factor of the first-order derivative is $p-2$ whereas it should be 1 for the standard equation of the associated Legendre polynomials. However, we still hope that we can reduce the equation to this by the appropriate transformation. We try

$$\Theta = (\sin\theta)^a H \,,$$

where a is to be determined. With this substitution

$$\frac{d^2 H}{d\theta^2} + (p - 2 + 2a)\cot\theta \frac{dH}{d\theta}$$

$$+ \left[\frac{A + a(a-1) + a(p-2)}{\sin^2\theta} - (B + a^2 + a(p-2)) \right] H = 0 \,.$$

Notice that we can now make the numerical coefficient in front of the first-order derivative equal to 1 if we set

$$a = -\frac{p-3}{2} \,.$$

Then

$$\frac{d^2 H}{d\theta^2} + \cot\theta\, \frac{dH}{d\theta} + \left[N(N+1) - \frac{m^2}{\sin^2\theta} \right] H = 0,$$

where

$$N = n + \frac{p-3}{2}, \quad m = -\left(n' + \frac{p-3}{2} \right).$$

This is exactly the equation for the associated Legendre polynomials. If you prefer, you can put it in the non-trigonometric form using the substitution $t = \cos\theta$:

$$(1-t^2)\frac{d^2 H}{dt^2} - 2t\, \frac{dH}{dt} + \left[N(N+1) - \frac{m^2}{1-t^2} \right] H = 0.$$

Therefore, we have found that

$$Y_n = (\sin\theta_{p-2})^{-\frac{p-3}{2}}\, P_{n+\frac{p-3}{2}}^{-n'-\frac{p-3}{2}} (\cos\theta_{p-2})\, \tilde{Y}_{n'}.$$

Notice the obvious inequality $n \geq n'$. Recursively

$$Y_n = Y_{n_1}(\phi) \prod_{j=2}^{p-2} (\sin\theta_j)^{\frac{1-j}{2}}\, P_{n_j+j-\frac{1}{2}}^{-n_j-1-j+\frac{1}{2}} (\cos\theta_j),$$

where $Y_{n_1}(\phi)$ are the spherical harmonics in two dimensions

$$Y_{n_1,1} = \frac{1}{\sqrt{\pi}} \cos(n_1\phi), \quad Y_{n_1,2} = \frac{1}{\sqrt{\pi}} \sin(n_1\phi),$$

and $0 \leq n_1 \leq n_2 \leq \cdots \leq n_{p-2}$ with $n_{p-2} = n$. It is more precise to indicate the spherical harmonics by $Y_{j n_1 n_2 \cdots n_{p-3} n}$, $j = 1, 2$. ∎

Bibliography

[1] M. Abramowitz and I.A. Stegun. *Handbook of Mathematical Functions: with formulas, graphs, and mathematical tables.* Dover Publications, 1965.

[2] N. I. Akhiezer. *Lectures on Integral Transforms.* American Mathematical Society, 1988.

[3] G. Arfken, H. Weber, and F. Harris. *Mathematical Methods for Physicists, Fifth Edition.* Academic Press, 2000.

[4] S. Bernstein. *Démonstration du théorème de Weierstrass fondé sur le calcul des probabilities.* Comm. Soc. Math. Kharkov **13** (1912) 1.

[5] F. Byron and R. Fuller. *Mathematics of Classical and Quantum Physics.* Dover Publications, 1992.

[6] C. F. Dunkl and Y. Xu. *Orthogonal Polynomials of Several Variables.* Encyclopedia of Mathematics and its Applications, volume 81. Cambridge University Press, 2001.

[7] G. Folland. *Fourier Analysis and its Applications.* American Mathematical Society, 2009.

[8] S. Friedberg, A. Insel, and L. Spence. *Linear Algebra.* Prentice Hall, 2002.

[9] H. Hochstadt. *The Functions of Mathematical Physics.* Dover Publications, 1987.

[10] J. Jackson. *Classical Electrodynamics, Third Edition.* Wiley, 1998.

[11] L. Landau and E. Lifshitz. *The Classical Theory of Fields, Fourth Edition.* Butterworth-Heinemann, 1980.

[12] L. Landau and E. Lifshitz. *Quantum Mechanics: Non-Relativistic Theory, Third Edition.* Butterworth-Heinemann, 1981.

[13] W. Miller, Jr. *Symmetry and Separation of Variables.* Encyclopedia of Mathematics and its Applications, volume 4. Addison-Wesley, 1977.

[14] J. Nunemacher. *The largest unit ball in any euclidean space.* Mathematics Magazine **59** (1986) 170.

[15] H. Royden and P. Fitzpatrick. *Real Analysis.* Prentice Hall, 2010.

[16] W. Rudin. *Principles of Mathematical Analysis.* McGraw-Hill, 1976.

[17] R. Shankar. *Principles of Quantum Mechanics.* Springer, 1994.

[18] J. Stewart. *Calculus: Early Transcendentals.* Brooks Cole, 2007.

[19] D. Singmaster. *On round pegs in square holes and square pegs in round holes.* Mathematics Magazine, **37** (1964) 335.

[20] W. Strauss. *Partial Differential Equations: An Introduction.* Wiley, 2007.

[21] J. D. Talman. *Special Functions: A Group Theoretic Approach.* W. A. Benjamin, Inc., 1968.

[22] N. J. Vilenkin. *Special Functions and the Theory of Group Representations.* American Mathematical Society, 1968.

Index

1-sphere, 23
2-sphere, 23, 92

angular momentum
    eigenvalues, 11, 12, 72
    operator, 10, 11

Bernoulli distribution, 53, 62
Bernstein polynomial, 49, 51, 53,
    55, 62
Bessel's inequality, 59
beta function, 120, 121

Cauchy product, 64
Cauchy sequence, 57, 60, 61
Cauchy-Schwarz inequality, 40, 61,
    89
Chebyshev polynomials, 85
Chebyshev's inequality, 54
circle, 124
closed set, 60
complete normed space, 58
complete set, 6
complete sets, 58
coordinate rotations, 14

del operator, 13
divergence theorem, 26
dot product, *see* inner product

electrostatic potential, 33
ellipse, 124
ellipsoid, 35, 124
Euclidean metric, 13
Euclidean space, 13
Euler's equation, 64, 74, 131
Euler's formula, 5

falling factorial, 46
Fourier series, 6

gamma function, 20, 120, 124
Gegenbauer polynomials, 99, 136
Gram-Schmidt process, 41
gravitational potential, 33
Green's functions, 110–115
Green's theorem, 73, 114

harmonic function, 33, 34, 116,
    119, 132, 135
harmonic polynomial, 68–75, 78
Hecke-Funk theorem, 102
Hilbert space, 58
homogeneous function, 116, 131
    degree, 116
homogeneous polynomial, 5, 12
homogeneous polynomials, 63–71
hyperellipsoid, 35
hypersphere, 35

inner product, 39

Jacobi polynomials, 99

Kronecker delta, 74

Laplace equation, 1, 33, 111, 113, 114
  in $\mathbb{E}^2$, 3
  in $\mathbb{E}^3$, 8
  in $\mathbb{E}^p$, 68
Laplace operator, 1, 5, 8, 13, 14, 30, 33
  in 2D spherical coord's, 2
  in 3D spherical coord's, 8
  in spherical coord's in $\mathbb{E}^p$, 33
Legendre polynomials, 78–105, 116
  addition theorem, 80
    in $\mathbb{E}^2$, 85
    in $\mathbb{E}^3$, 84
  associated, 83, 117, 137, 138
  differential equation, 96
  generating function, 115
  integral representation, 104
  norm, 93
  orthogonality, 94
  parity, 80
  recurrence relation, 97
  Rodrigues formula, 95, 100
linear combination, 58

mean value theorem, 31
method of images, 114
metric tensor, 34, 123

n choose k, 46
nabla, *see* del operator

norm, 42

orthogonal matrices, 14
orthogonality, 5, 40
orthonormality, 58

p-ball, 21, 34, 36, 124, 126
  closed, 20
  open, 20
p-cube, 36, 126
p-ellipsoid, 35, 124
p-sphere, 20–22, 25, 34, 35, 92, 93, 117
  infinitesimal arc length, 26
  surface area, 21
Parseval's equality, 60
Pythagorean theorem, 25

random variables, 53
recurrence formula, 42
Rodrigues formula, 48

Schrödinger equation, 2, 6
separation of variables
  in three dimensions, 8
  in two dimensions, 3, 7
Singmaster's theorem, 36, 126
spherical coordinates
  in $\mathbb{E}^p$, 18
  in $\mathbb{E}^2$, 2, 15
  in $\mathbb{E}^3$, 8, 16
  in $\mathbb{E}^4$, 17
  inverse relations, 19
spherical harmonics, 71–117
  closedness, 106
  completeness, 108

expanded in Legendre polyn.,
    87
general definition, 71
in $\mathbb{E}^2$ (on S^1), 4
orthogonality, 72
parity, 72
spherical Laplace operator, 33
    eigenfunctions, 71
Strirling's approximation formula,
    124, 127
substitution of variables, 124

surface harmonics, 84

tesseral harmonics, 84
triangle inequality, 40

weak law of large numbers, 54
Weierstrass approximation theo-
    rem, 50
weight function, 40

zonal harmonics, 84

Printed in the United States
By Bookmasters